SOTHIS PRESS

Sothis Press aims to completely reimagine the public domain book with an emphasis on universal access to captivatingly designed print editions. We believe that technology will enable every single book ever published to be perpetually accessible in beautiful print form. Sothis is beauty in ubiquity.

www.sothis.press
1 Waterhouse Square, London, ECIN 2ST

The Waters Above the Firmament, or the Earth's Annular System

The Mosaic Record Scientifically Explained

Isaac Newton Vail

SOTHIS PRESS

ISBN 978-1-0368-0396-4

The Waters Above the Firmament, or the Earth's Annular System is in the public domain.

This edition is published by Sothis Press The company is registered in England and Wales. The Company Number is 14900389.

The Earth's Annular System

OR

The Waters Above the Firmament

THE WORLD RECORD SCIENTIFICALLY EXPLAINED

By ISAAC N. VAIL

AUTHOR OF "THE MISREAD RECORD," "EDEN'S FLAMING SWORD,"
"ALASKA, LAND OF THE NUGGET, WHY?" ETC.

Fourth Edition

"Yet I doubt not through the ages one increasing purpose runs,
And the thoughts of men are widened with the process of the suns."
—*Tennyson*

Pasadena, California:
THE ANNULAR WORLD COMPANY
411 Kensington Place
1912.

PREFACE.

On the 26th of January, 1912, Isaac N. Vail, the author of this book and the originator of the Annular Theory of Evolution, died suddenly at his home at 411 Kensington Place, Pasadena, Cal.

About the first of December, 1911, the previous edition having been exhausted, he had made arrangements for a new edition of this, his largest published work. Only about half the plates were finished at the time of his death, so that it became the duty of his daughters, Alice Vail Holloway and Lydia C. Vail, in whose possession he left all his published and unpublished writings, to carry on the work. This volume now goes before the public just as the author left it.

Isaac N. Vail was essentially a student, modest and retiring in his nature, and with but little of the aggressiveness of either the propagandist or the man of business that would have pushed his views before the world. In his later years he found increasing contentment in searching out material to be used by those who would continue the work after he had left it.

No one realized so fully as he the far-reaching results and revolutionary effects of annular thought upon almost all departments of science and philosophy. With this realization, and with a confidence and patience born of a knowledge of the truth, he was willing to wait till the world was better prepared to accept his ideas.

He knew that in the light of new discoveries old theories must fail, and that in time the Annular Theory of Evolution must gain a sure foothold in the young and vigorous minds of coming generations.

To the end of carrying toward completion the work begun by our father, our lives will be devoted. and as the demand arises we hope to give to the public further evidence in support of annular evolution.

We undertake this work not only with the feeling that we are fulfilling a sacred trust, but with a love of the work for itself, and with a profound conviction of its importance to the progress of science.

While we may be able from time to time to present to the world even vast accumulations of evidence gleaned from the fields of geology and especially mythology, we can do but little more than suggest to more able and scholarly minds than our own the work to be done in the fields of Philology and Biology.

A new science of the origin and growth of language will have to be written. The Darwinian theory of evolution will be found insufficient and will have to be supplemented by the Annular Theory of organic evolution. Following this must come a new Ethnology and a modified Sociology.

It is with the hope that we may gain the attention of those who are able to do better work than we, that we now send forth this volume.

ALICE VAIL HOLLOWAY.
LYDIA C. VAIL.

Pasadena, Cal., April, 1912.

INTRODUCTION BY THE AUTHOR.

The first edition of the "Waters Above the Firmament," or "The Earth's Annular System," was published in pamphlet form in 1874. The chief effort of the author at that time was to show that the Deluge of Noah, and all the "Ice Ages" were caused by the progressive and successive collapse of great world-canopies of aqueous vapor, which were the last remnants of a Saturn-like Ring System, or a Jupiter-like "Cloud-Ocean," sent to the terrestrial skies when the earth was in its molten stage. In that edition it was also shown that earth-canopies were competent to produce all the tropic eras the earth ever saw. It was also pointed out that world-canopies trending poleward tended to mass themselves in the polar heavens, and fall in the Arctic and Antarctic regions as vast avalanches of snow.

More than forty years of research along these lines have more than justified the claims originally set forth. As a practical geologist I have gathered facts from the earth-record directly, and I am bold to aver that these facts unquestionably demand a revision of geologic thought, as it stands to-day. The tremendous truths of world-evolution shine all the more perfectly under the arc-light of annular earth-building.

It was this independent research in a very wide field of thought that led me to enlarge the pamphlet of 1874 to a book of 400 pages in 1885; and again it was revised and enlarged in 1902; and I have been greatly encouraged by the fact that this last edition is now used in some of the colleges, and in at least two universities as an educator.

When the first volume was published in 1874 it was a rare thing to meet with a scientist who would admit that the earth had a ring system; to-day it is as rare to meet with one who does not concede the great fact, and the great problem is resolving itself into this form: How did the earth's rings fall back to the surface of the planet? I have attempted to answer this question in the following pages, and as this fourth edition is being prepared for the press I have before me more than a thousand letters from many parts of the thinking world. The great mass of them concedes the logic of my contentions. But three of their writers have taken a variant view.

Annular earth-evolution, during the last two decades, has assumed a stage of supreme importance. During this period I discovered many old thought petrifactions in the oldest world-literature of the races, which seemed to point directly to the reign and fall of an ancient earth-canopy. Keeping this central fact in view, during the last ten years, I have secured from the old beds of fossil thought the most indubitable evidence that when the oldest records were inscribed the last remnants of the earth's annular system lingered prominently in the terrestrial skies. When the mature and reliable judgment of this generation is pronounced it will be to establish the great fact that the geologic record is a time-written history of the reign and fall of earth-rings, as reflected in the unmistakable reign and fall of canopies.

The "Old School-Geologist" will likely be the last person to admit that the earth's aqueous strata have largely fallen into place as we find them to-day, as the giant wreck of slowly declining annular matter; but when it shall have become apparent, that an annular system does not necessarily fall as a sudden titanic world-collision, but as continuous world-showers of dust and other meteoric matter, and floods of watery vapor, and snow, through the "Ages," the great mass of the thinking world will readily admit the logical record which declares that not only the great mass of the mineral crust of the earth has to a vast extent been built up, sometimes very slowly, and again very rapidly, as annular material, but it will also admit that the very life-succession that characterizes the record, can only be explained by the logic of annular-canopy world-evolution.

It was not until after this century began that the illustrions W. F. Warren, Dean of Boston University, kindly called my attention to the fact that the immortal Kant more than 150 years ago suggested that the fall of waters from an earth-ring might have caused the Deluge of Noah. I desire to concede to that great philosopher all the glory his suggestion merits. To this end I desire that the students of annular and canopy philosophy should read all that he seems to have written on this subject, and especially his reasons for not advocating the thought. It is found in "Kant's Cosmology," by Hattie; pp. 229 to 231.

In the more than forty years of careful examination of the fossil records I have found an amazing amount of testimony that gives support to this contention; but the most astounding part of this testimony is the strange fact that nothing in the old records has been found logically arrayed against it. There is a natural scheme found prominent in all earth-building that harmonizes and dovetails into the establishment of a law of progression, and succession which nothing short of the decline of rings and the reign of earth-canopies successively can satisfactorily explain.

Life-mutation is unquestionably so linked to the history of our planet as to show plainly the frequent passing away of old world-conditions, with catastrophe more or less striking, and the advent of new life with new conditions. Plainly new life-germs have settled down on the ruins of the old. In other words life-mutation is fossilized in the rock record and tells its own tale of world-changes. The logic of Nature confronts us with the grand plan of life-stages, the succession of ages. The system is seen in the great world-scheme, and in the last analysis of the earth-problem we will be taught that if the earth never had had an annular system it would never have had a succession of "ages," except as accidental stages.

The succession of tropic conditions which the earth-record affirms, also affirms canopy-succession, and canopy-succession means progressive ring declension. We learn that all earth progress accords with the fact that gravital law determined the grand scheme when it determined the progressive movement of the fiery exhalations that the molten earth sent up into the cold of the skies, through all degrees of temperature, and thus afforded an opportunity for the formation of those compounds found so abundantly in the earth's crust.

The reader's attention is called to the cut facing page 175 of this volume, and which was first published in the 1902 edition of this work. Canopies must float pole-ward in falling, and leave such features persistently in the polar skies. Just before the close of last century I found this polar picture fossilized in the oldest thought of many peoples. The scholars of the world must know that the ancient North World was a point of supreme regard for all humanity. I repeat that canopy vapors massing themselves in the polar skies and plunging to the earth, must have left some "openings," or "breaches," literally "star-isles" in the polar heavens. I have found such sky-rifts as thought-petrifactions with almost all the ancient races, and need only call attention to the Bel-Peor, or "Shining Hole," whose image was a "hole in the wall." (Ez. 8: 7.)

It is plain that if such features are now found as petrifactions in the oldest thought, there can be no questioning the claim that the earth had an annular system, and that its remnants of vapor and mineral dust lingered in the terrestrial heavens long after the advent of civilized man. This thought is more fully elaborated in my book "The Misread Record," or "The Deluge and Its Cause." (1905).

ISAAC N. VAIL.

PASADENA, CAL., Eleventh month 11th, 1911.

LIST OF ILLUSTRATIONS.

CONTENTS.

The Earth's Annular System.

CHAPTER I.

ALL WORLDS MADE ALIKE.

To endeavor to prove the truth of the theory that supposes this earth to have been, from the close of the igneous era till the close of the antediluvian period, surrounded by an annular system, seems to me, since I have been so long gathering in the fund of evidence, like trying to establish a self-evident truth; yet, since geologic science has been pursued by a host of honest and indefatigable workers, with ideas at variance with this claim, and since established theories are not expected to be abandoned abruptly, it is plain, that as the annular theory demands a general and thorough review of the geologic record, as now interpreted, a great effort will be necessary to bring it within the purview and consideration of science.

The geological "column" reveals many facts that have not yet been recognized by investigators. It would be strange, indeed, if frail and erring man should have erected a faultless fabric out of the crude materials supplied; strange, indeed, if the "records" have in all cases been interpreted without fault; strange, indeed, if the edifice will not some time have to be taken down and rebuilt, as it has been at different times. The time was, when investigators were "few and far between." Thousands of eyes now run over the field where trod the investigators of fifty or a hundred years ago, and the day is past when a geologist can sit in his studio and frame a theory for the great mass of thinkers. The true theory was written by the hand of the great Master-builder, and it must be read on the spot, where the wondrous mystery lies.

It is with no desire to find fault with, or undervalue the work of the noble band of ardent workers now in the field, that I advance the claim that we have greatly misunderstood and misinterpreted the fascinating volume whose time-stamped pages are unfolding to our view. We have misapprehended the rock-engraved hieroglyphics from the very first rude lines, traced on the archæan piles.

It was a sublime conception of Prof. Winchell that represented all the waters of the terrestrial oceans as held in suspension on the outskirts of the primeval atmosphere by the inveterate heat of the igneous earth.

The reader, no doubt, remembers his glowing description of the Titanic contest between the powers of Vulcan and Neptune. How the waters on high descended, while yet the earth was a hot and seething mass, and were again and again flung into space by the irritated fires; till, finally, worried by the eternal attacks of Neptune, the fires grew tame, and the oceans of vapor settled upon the earth. Thus is portrayed the mistaken idea now universally prevalent, that from that period, beginning as soon as the waters could remain upon the earth, these having all descended upon it, worked as the universal ocean now does in building up the aqueous crusts; that the work of denudation and the distribution of detrial matter was participated in by the entire ocean as it now exists, from the very period when the internal fires grew quiet and permitted the waters to remain upon the surface of the planet.

Now is this claim philosophic? Did the oceans all descend at that time? I try to settle this point by the test of philosophic law; for, here is the foundation stone of the geologic edifice men have built, and coming scientists will lift it from its bed and relay it. I ask geologists to critically examine this point and see whether it be not a fact that only a small part of the oceans fell at the time alluded to, and that the remainder continued to revolve about the earth for immeasurable time, as an annular system, or a belt system like that of the planet Jupiter. Now if the oceans all fell to the earth at the close of the igneous era, then the current theory of crust evolution is correct; but if not, it is incorrect.

It will be shown in the following pages that the terrestrial waters did not all fall at that time; that physical law demands that they should not. This is susceptible of the clearest demonstration. The importance of this question cannot easily be overestimated; and yet, the first thought may be, "Of what value is it?"

The reader who patiently reviews this problem will see that a more important one has never come before man for his consideration.

In the first place, let me press this question: Is there any thing unreasonable or unphilosophic in the claim, that the aqueous vapors, kept away from the molten earth by the repelling force of heat, were necessarily whirled into independent revolution about the central fiery orb? Since we see at least two giant planets in the solar system attended by such revolving vapors, is it not a reasonable claim? It is conceded that Jupiter's belts are aqueous vapors.

These make a complete revolution in about ten hours, and it is claimed by eminent astronomers that they are held away from the planet by his own native heat.

Well, suppose this heat were suddenly removed? The Jovine atmosphere would contract and what would become of Jupiter's moving belts? They are so many tons of moving matter, possessing so many tons of moving energy, and every one must see that that energy would prolong their stay in Jupiter's firmament. It is evident that revolving vapors would be no more likely to fall immediately upon the withdrawal of heat, than a revolving moon in the same situation; and if their moving energy was great enough, it is plain that Jupiter's belts would continue to revolve independently about him after he had cooled down. Now, since the equatorial belts of both Jupiter and Saturn move more rapidly than the polar, they must be moving independently of each other and also independently of the bodies of those planets. That is, they do not move in those planets' atmosphere, but are revolving about them in their own independent orbits. Then Jupiter's belts do possess energy sufficient to insure their continuance in a belted or annular system revolving about him, for unknown time. This feature of the question will be fully elaborated in another chapter.

We know that the terrestrial waters, like Jupiter's, were at one time kept away from the surface of our planet, and we know, too, that in the revolving mass, a moving energy was imparted to these also, and that that energy must have prolonged their stay in the terrestrial firmament, after the earth cooled down. One must see that on the very threshold of this investigation, my claim that the earth's oceans did not all return to the earth at the close of the igneous era, is a reasonable and philosophic one. I might almost say a necessary one.

No geologist, astronomer, or physicist, will, I presume, for a moment doubt the now firmly established conclusion, that the earth was at one time in an igneous-fluid state; and also that while it remained in that fiery condition, all its waters and whatever else that was vaporized and sublimed by the inveterate heat, such as the less refractory minerals and metals in the boiling mass, were driven away from its surface and hindered from falling upon it by the repelling energy of heat. A failure to follow this conclusion, and the consequences necessarily flowing from this primitive condition of our planet, has involved us in a maze of difficulty and error. A failure to comprehend many of the legitimate consequences of the measureless force employed, every pound of which must have been conserved in after-effects, has immeasurably checked the solution of some of the grandest problems of Nature.

Let us now begin at the very foundation of this physical problem, and critically examine every step of our progress. We will reject every link of evidence that will not bear the test of scientific scrutiny. I must ask the reader to patiently follow me in the line of argument I am about to pursue, for it requires a round of investigation that few men will at first appreciate.

Our foundation is the molten, or igneous world. The vaporized water, mineral and metallic elements repelled from it, existed as a great vaporous atmosphere, that rotated with the earth, just as our atmosphere now does. If the earth then rotated once in every twenty-four hours, the atmosphere turned with it in the same time. If it rotated in the short space of about three hours, as claimed by Proctor and other eminent astronomers, the great primeval atmosphere rotated with it in three hours. Does not this postulate demand unconditional assent from all men? Will any fair reasoner claim that I assume here what is not self-evident? A little thought will induce the most incredulous to admit that my claim here made is just and necessarily true. Let us remember, then, that the primeval atmosphere rotated with the earth—in the same time, no matter how long or short that period was. Then the question is at once reduced to this:—When did those vapors and other material constituting that atmosphere return to the earth? For they *have* returned. The question, *how* did they return, is also a legitimate one, and will receive due consideration.

They returned or fell to the earth, either immediately after it cooled down and the heat ceased to repel them, or they continued for a time to revolve around it. If some of those vapors fell at the close of the igneous era, then a part of them continued to revolve. As before intimated, the science of Geology has been built entire upon the former supposition, and the annular theory is planted upon the latter. Witnesses must determine which of these foundations is false; with a reasonable probability in the truth of the latter, as attested by the Jovial and Saturnian belted or annular systems; and the improbability that the potential energy stored up in the rotating mass of vapors during its repulsion by heat, would all be expended in their decline in the period between azoic and paleozoic time.

The most eminent scientists agree that the great mass of swaddling vapors in the primitive atmosphere were driven at least 200,000 miles from the earth. Others claim that the earth's vaporous atmosphere during the igneous era, embraced the orbit of the moon within its boundaries. It must be remembered that all the carbon in the great casement, of aqueous rocks, the vast oceans of oxygen now containedin the silicates, sulphates, carbonates, and oxides of the crust, as well as the nitrogen and hydrogen, in numerous compounds, enormously swelled its volume, so that a modern chemist speaking from his laboratory, makes the claim that if that atmosphere pressed on the earth in proportion to its depth as ours does to-day, unaffected by repelling heat, it would be equal to a column of mercury more than 22,000 inches high. I believe it was M. Figuier that first advanced the idea that this atmosphere extended to the moon, and others would extend it still farther. This, of course, is understood to be its extent at the close of the igneous era, and before the aqueous beds were laid down. Now, however conditioned the atmosphere was at that time, one thing is very evident, it was one of vast extent. If I should take advantage of these claims and base my calculations upon an atmosphere 200,000 or 240,000 miles deep, it would greatly aid me, and make my conclusions much more apparent and conclusive. But to be sure that we are moving entirely within philosophic bounds, and to give no possible opportunity for an opposer to claim that I strain any point or take undue advantage of extravagant admissions of men of science, I will not claim

240,000 nor 200,000 miles as the atmosphere's depth, but will base my calculations on a depth of only 100,-000 miles. This is amply sufficient for us, and with this depth it is easy to prove beyond a doubt, that a mighty fund of vapors continued to revolve for unknown time about the earth.

Again, it is to-day a favorite theme of astronomers that, during the igneous era, the earth rotated in a period of only three or four hours. If this be true, the probability that the matter in the primeval atmosphere was whirled into belts or rings is increased from six to eight fold. It seems scarcely needful for me to say, that astronomers came to this conclusion by a legitimate process of philosophic deduction. It must be evident that this rate of rotation would be of great advantage to us in establishing annular conditions; for, almost every school-boy has learned that if the earth should rotate more than seventeen times as rapidly as it now does, the oceans at the equator would be whirled into space, and made to revolve around it. Then, a rotation in three hours, or eight times as rapidly as at present, would whirl matter already floating in the atmosphere to a greater height and increase annular tendency in the same proportion. However, we will decline to make use of this advantage, and use only that rate of rotation that every one knows to be correct, viz:—one revolution in 24 hours.

Here, then, we have true philosophic data which all men will certainly admit to be fair; and upon which all may proceed to erect the annular theory, and we will endeavor to square every timber in the edifice by one unvarying rule:—Philosophic Law. If we succeed with these data to start with, men of science may multiply its certainty by at least twelve, for their own satisfaction. The data then are:—a primeval atmosphere admitted on all hands to be 100,000 miles deep, and a known velocity of rotation of once in 24 hours.

With this rate of rotation, we also know that the velocity of any point on the equator of the earth was about 1,000 miles per hour, while the equatorial periphery of the great vaporous atmosphere moved with an actual velocity of more than 25,000 miles per hour. This, the most ordinary mind can determine; but as we are searching for facts that any child who may peruse these pages may understand, I will give the simple calculation here.

If the atmosphere were 100,000 miles deep, and the earth 8,000 miles in diameter approximately, the diameter of the sphere would be 208,000 miles, and the circumference a little more than three times that or about 624,000—the space that any point in the outer boundary of the atmosphere would move through in 24 hours,—and of course $\frac{1}{24}$ of that distance in one hour, or 26,000 miles (I will give 1,000 miles to the other side out of pure liberality).

The simple conclusion drawn from this is, as any one can see, that a ton of matter at or near the equator of the earth would have a momentum of 1,000 tons, in the rotating mass, while a ton of vapor or any other matter on the peripheral boundary of the atmosphere, would have a moving energy of 25,000 tons.

Suppose the former were placed ten miles above the surface of the earth, and the latter brought down to the same position; the former with a velocity of 1,000 miles per hour would immediately fall to the earth, while the latter would rise, and revolve around the earth as a satellite, as can be readily proved by a simple calculation. The mass possessing 25,000 tons of moving energy must lose 8,000 tons of that moving force before it would, or could reach the earth; for as I have before stated, it is a well-known fact that any body moving around the earth at a rate of more than 17,000 miles per hour, can never fall to its surface, and a ton moving at that rate would possess 17,000 tons of momentum, and it becomes a known fact that if that momentum were increased to 25,000 tons, or a velocity 25,000 miles per hour, it would rise and revolve in its appropriate orbit about the earth, and never until its velocity became diminished to about 17,000 miles per hour could it reach the surface of the earth. Now it could make no difference whether a body be a ton of stone or a ton of aqueous vapor, it would continue to move around the earth so long as the centrifugal exceeded the gravital force. Hence it is evident that upon the data assumed above, of an atmosphere less than half so extensive, as scientists assumed, and with a radial velocity more than six times less than they claim for the mass, the centrifugal force of a vast portion of the aqueous vapors and other matter in the primitive atmosphere was such as to effectually hinder their fall to the earth, as the latter cooled down and the vapors condensed. It is also evident that the matter in the lower regions of the atmosphere would fall on the withdrawal of terrestrial heat, and it is an easy thing to ascertain the line, or height in the atmosphere beneath which all vapors upon condensing would fall, on account of insufficient centrifugal force or moving energy to keep them there, and all vapors beyond which would remain there because of insufficient gravital force to bring them down.

What, then, must have been the condition of those materials that formed the upper and outer stratum of that great atmosphere after the earth became cool and the atmosphere shrank to near its present dimensions, and all the aqueous matter, etc., to the height of 20,000 or 30,000 miles had fallen to the earth? These must have been vast oceans of clouds possessing a velocity that prevented their descent, and which continued to move around the earth; that is, the earth had an annular system. If any criticism can shake this conclusion, there is nothing in law! One would suppose that this is all-sufficient to settle the question forever, that the oceans did not all fall to the earth at the close of the igneous era, but that such as existed when they had not centrifugal force sufficient to retain them on high, did fall; but I will not put this conclusion aside until I have shown still further the impregnable grounds upon which it is based. It is easy to demonstrate by a mathematical calculation that the above depth of atmosphere and rate of rotation are much greater than that which was actually necessary to produce annular formation about the earth.

The analytical expression used by mathematicians to represent the whole force of gravity at the earth's equator is $g + \frac{c^2}{D}$, where g is the visible force of gravity, or the space a body will fall at the equator during the first second of time; c is the chord of an arc over which a revolving body moves in one second, and D the diameter of the orbit of which c, or the arc, is a part, and $\frac{c^2}{D}$ is the centrifugal force or the part of gravity destroyed by rotation, or movement in an orbit. It is evident that the arc c or the space passed over by the moving body in one second, will be practically equal to the chord of the same arc, and I will therefore use it as such; that is, as a straight line. Now, as $\frac{c^2}{D}$ is the centrifugal force, and g the gravital or centripetal force, when these forces are equal, and the body neither falls nor rises, but moves on continually in its orbit, $g = \frac{c^2}{D}$.

Now there are 86,164 seconds in one complete rotation of the earth, and the circumference of the earth is $D \times 3.1416$ nearly; and this divided by 86,164, number of seconds in one rotation, gives the length of the arc c, or the distance any point on the equator moves in one second of time; in other words, the rate of motion. But when $g = \frac{c^2}{D}$ it is evident that $gD = c^2$ or $c = \sqrt{g D}$ and as often as c, the distance a body moves in one second, is contained in the whole circumference, so many seconds are there in one revolution; that is, $D \times 3.1416$ divided by c or its equal, $\sqrt{gD}$, thus: $\frac{D \times 3.1416}{\sqrt{gD}}$ = number of seconds in one revolution when $g = \frac{c^2}{D}$ or when the earth rotates so rapidly that the centrifugal force on the equator equals gravity. Then we evidently have

$$\frac{D \times 3.1416}{\sqrt{gD}} = \frac{7925 \times 5280 \times 3.1416}{\sqrt{16.076 \times 7925 \times 5280}}^{*} =$$

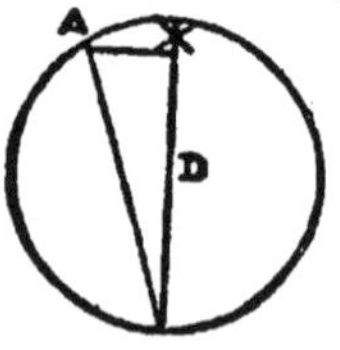

FIG. I.

* 7925 × 5280 = number of feet in the earth's diameter, and 16,076 = g = distance a body falls at the equator during the first second. Let D be the earth's equatorial diameter (7925 miles), and X the versed sine of the arc or distance a point on the equator moves in one second; A X is the chord of the arc, and practically equal to the chord itself, where so small a portion of time is considered.

5069 seconds $= 1^h, 24^m, 29^s$, or the time in which a ton of matter would have to revolve about the earth just at its surface at the equator, so that it would neither rise nor fall, and when, if its velocity were increased, it would move away from the earth, and in another orbit. Now this velocity is 17 times the present velocity of the earth's rotation, or about 17,000 miles per hour. Hence, we have an absolute demonstration that any body in our present atmosphere or in the great primeval atmosphere, or at any point above the earth, moving at the rate of 25,000 or 20,000 or even 17,500 miles per hour around it, could not fall to its surface! But vast quantities of primeval vapors did move with this velocity according to our assumed data, which data we have no reason to dispute, and therefore we are abundantly justified in the claim that the earth for unknown time was accompanied with an annular system, and the geological record has been misinterpreted, and must be reviewed, and geological theories remodeled.

The foregoing calculations, it might seem, are all-sufficient to establish the fact of annular formation about the primitive earth. But this formation, once effected, demands a permanency of existence, which an immensity of time only can effect. Rings once formed about the earth after the lapse of countless millions of years, cannot collapse in a day. They must lose their momentum with a steadiness as invariable as the flood of ages. It would be as unreasonable to suppose the earth's present satellite would in an hour break loose from its anchorage, and descend to the earth, as to suppose that one of its rings could do the same thing. Then with the primitive earth surrounded with a ring system whose longevity could be counted only by geologic ages, are we for a moment to suppose that the aqueous strata-formation only began after that system had fallen? Is it not more reasonable to suppose that the aqueous strata began to form as the vapors began to descend, and that the latter continued their decline through all geologic time? What is there unreasonable in the claim?

Rings of aqueous vapor, however associated with mineral and metallic matter, must follow in all respects the same laws as a moon or planetary satellite in their motions around their primary. There is a law well known to the mathematical world, called "Kepler's Third Law." Let us bring it into use. By it we can readily demonstrate not only that the primitive distillations, repelled from the fiery sphere, were thrown into a ring-system, but by it we can also readily show how far above the earth's surface they must have revolved about it. This law may be stated thus: The squares of the periodic times of revolving satellites are proportional to the cubes of their mean distances from the primary around which they move. This is of universal application whatever be the shape or constitution of the satellites, as all must know. Then if we take the cube of the radius of the moon's orbit, which is sixty times the equatorial radius of the earth, and divide it by the square of the time of its revolution in seconds, it must be equal to the cube of the orbital radius of a ring of any kind of matter revolving about the earth, divided by the square of the time of its revolution in seconds.

As before stated, the primeval atmosphere in which the matter distilled from the igneous earth existed, and out of which matter all terrestrial rings must have been formed, rotated with the earth; and we have assumed this rotation to be once in twenty-four hours, which the reader will readily grant. Then it must be seen that we have three known terms of a proportion to find the fourth. This fourth term is readily found, and is the actual distance of any terrestrial ring from the earth's center. Put this unknown quantity = *R* and we will have the following easy calculation. The time of the primitive atmosphere's rotaιion = 86,164 seconds, moon's time 2,360,608 seconds, and we have the following equation:—

$$\frac{R^3}{(86{,}164)^2} = \frac{60^3}{(2{,}360{,}608)^2}$$

developing and reducing by simple calculation or more readily by logarithms, we will find R^3 = 279.725264 and R = 6.54 times the equatorial radius of the earth, or the semi-diameter of a ring revolving about the earth once in twenty-four hours. In other words, vapors, of whatever kind, in the primitive atmosphere, at the height or distance of 26,000 miles from the earth's center, or a little more than 22,000 miles from its surface, possessed all the independent energy of a revolving satellite; and all vapors farther off possessed still greater momentum, and those nearer the earth did not possess the energy of a satellite, and fell to the earth as it cooled down, leaving the more distant matter moving independently about it. Is there anything wrong with this demonstration? Thus "Kepler's Third Law" establishes the truth of the annular theory, or proves itself to be of no value at all!

Then we must see that we need no atmosphere 240,000 miles in depth, nor 100,000, nor even 22,000 miles in order to show that annular formation was an absolute necessity in the evolution of the earth. Every mile added to this paltry depth, adds to the certainty of the fact. Did the earth then rotate in 86,164 seconds, or did it rotate in half that time? Every second of diminution adds to the certainty of the fact. How can we escape this conclusion? Thus is rendered plain and irrefutable the claim that the lower part of the great aqueous atmosphere, upon contraction and condensation resulting from the loss of terrestrial heat, fell away from the upper part, simply because the latter (like the rim of a great revolving wheel) moved so rapidly it could not descend, but continued to revolve about the earth until it lost so much of its independent inertia, as to permit it to descend, as will be shown in its proper place.

Proceeding thus from the known condition of the primitive earth along a track, every step of which is known, we have by adhering to strict philosophic demands, laid the foundation of a theory that no man can shake. The reader will from this time observe, that the fabric built upon this foundation, is not an obelisk, but a pyramid, whose successive stages add permanence to the adamantine sills upon which it stands.

Let us look back upon the ground over which we have passed. We see a fiery globe rolling through space with a vast and heavy atmosphere, rotating so rapidly that its outskirts are unavoidably made to assume such a velocity as to prevent them from falling. The earth was then a glowing sun, or a gleaming star, as analogy seems to prove. But when this earth from "its inmost bosom burned," when its oceans of molten minerals beat upon a seething coast, when its rivers were fluid fire, and its fountains dashing flames, when its "clouds by fiery tempests driven," dropped their steaming floods, an energy potential was stored up in the mighty upper deep—a vast abyss that literally built the aqueous world in after times.

We can easily imagine a world and its atmosphere turning so slowly that the vapors would fall immediately after it cooled down, leaving the heavens clear, and a vast universal ocean washing it. But no such conditions have ever existed on the earth that could bring these things to pass. Every mathematician must know full well, that the rotation of such a mass once in twenty-four hours, would inevitably separate the upper vapors from the lower, leaving the upper far above the atmosphere or terrestrial firmament, obeying the demands of inexorable law. And when investigators recognize this fact, as it stands to-day demanding a respectful consideration, then, and not till then, will they be able to unlock some of the most perplexing questions of science, which now defy explanation. It is the Philosopher's key to "nature's vast cathedral." I dare not now point out the grand avenues of thought which it opens; but time will make all things visible. I almost said, "all things new," not only in physics, but also in metaphysics! All I ask of the reader of these pages is implicit recognition of Law, in this field of labor so near the Great Fountain of Truth. The moment we leave it, we land in shadow and darkness. To propagate and teach *one* error hides a multitude of truths. An error taught in the name of science is a pernicious falsehood. We must, sooner or later, acknowledge the declaration of the missed and lamented Agassiz: "A physical fact is as sacred as a moral principle;" for a physical fact ignored sends violated impulses through the nerve centers of society; and their impress is traced in imperishable lines, as by a hand unseen. But one physical fact stands out prominent in the universe, viz: Annular formation is a necessity in the evolution of worlds from their primitive state ! !

It must be plain from the foregoing that if geologists had followed the grand train of philosophic events resulting from the igneous fluidity of the burning earth in primitive times, they must have long since concluded that much of the aqueous crust of the earth once existed with the revolving vapors, as infinitesimal particles, or tellurio-cosmic dust, in the ring-system. It must be so; from the very nature of that great sublimation of terrestrial elements, we are forced to this conelusion. And now if men of science will but open their eyes and look, they must see it. Let them follow this conclusion to its legitimate end, and they must see with the keenest regret, the fruitless toil of centuries. No, not fruitless! They have sailed along the very boundary of this field of investigation, and its explored avenues have yielded returns that will aid in the new realm of thought. The glories of a brighter day are dawning in mankind's sky, when all men must see more nearly eye to eye.

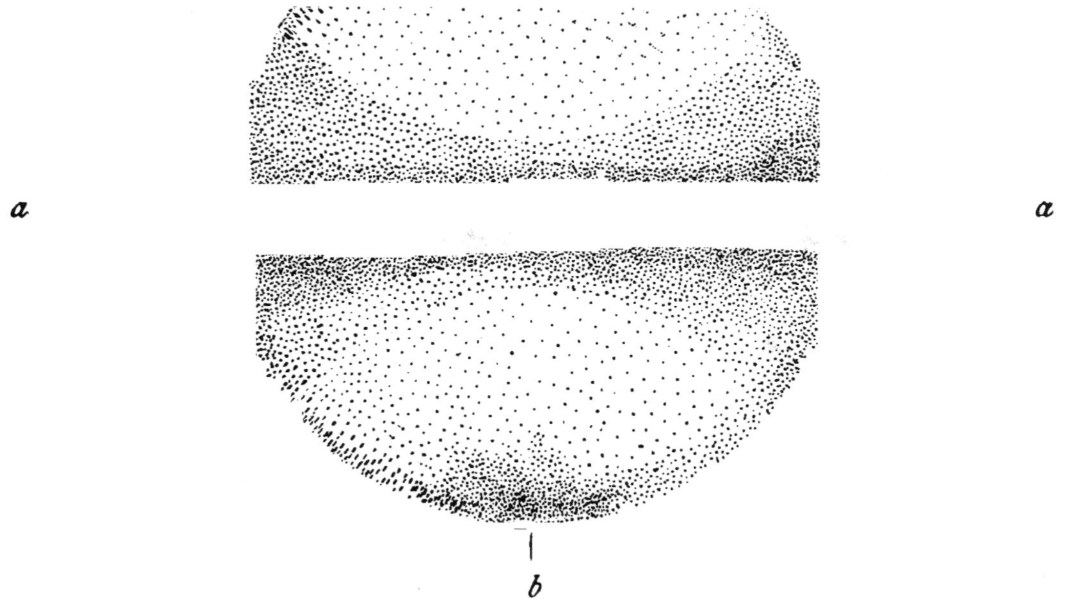

Fig. 1. EARLY STAGE OF RING FORMATION.

During the igneous era the rotation of the molten earth, supplemented by the expelling fires, carried all vapors, whether aqueous or metallic, to the terrestrial skies. Those vapors took the direction along which the combined forces (heat and centrifugal energy) were strongest, i.e., to the equatorial heavens. In that day the young earth, could it have been viewed from the moon, would have presented approximately the form shown in figure 1.

The incipient rings are seen forming over the equator (a, a) and all vapors lifted into the polar skies (b, b) would, by their forced rotation, be carried toward the equator, as water is thrown from a turning wheel

CHAPTER II.

SOME GENERAL CONSIDERATIONS.

To those who are able to follow physical causes to their legitimate and necessary effects, it will not be difficult to satisfactorily explain the mode of ring-formation about the primitive earth. The grand and stupendous changes that have recorded their way-marks, are the guide-posts of the investigator, pointing unmistakably to unavoidable conclusions. Being the physical means by which an incomprehensible Planner and Architect has completed a beautiful world for the habitation of man, according to unchanging law, if we follow philosophically the unmistakably known conditions stated in the previous chapter—i.e., those flowing from a state of igneous fusion, we cannot draw erroneous conclusions. Then let us be careful that no error enters the threshold of our work as we pass from the known to the unknown.

We all know with absolute certainty that when the earth was in a state of fiery fluidity, the entire oceans of water now on its surface were held suspended at a great distance from it. Not one drop of the mighty waters now surging against the shores of the earth, could have remained for an instant on its surface, nor in its flaming firmament. This we can all see so plainly that no one can entertain a doubt upon it. There is another known condition that will greatly aid us in this argument, when properly understood: the evident disposition of the cooling and contracting earth to absorb or draw the waters into its rock-forming crust, as they fell to its surface. It is not difficult to determine how much water or moisture a heated stone of any kind will absorb as it cools, so that we can give a reasonable and just approximation of the amount of water existing in the various rock formations of the globe. Many investigators have worked upon this question, and it has been found by analysis and experiment that even the solid lime-rock contains from one-half to one per cent. of water. Coals vary from five to twelve per cent. in the amount of water they contain. Iron ores from one to six per cent. Some sand-stones contain from six to ten per cent., clays much more, while granite, mica, feld-spar, and even solid quartz crystals, contain moisture in different amounts.

In our deepest caves and mines, some of them far below the ocean's level, water is found running down and into the bosom of the earth. This proves that the solid crust is still imbibing the waters of the ocean with a quenchless thirst, and that coming ages must diminish its volume and depth. If no more than a single barrel of water is absorbed in a day, a sufficient number of days will cause the solid earth to appropriate all the waters on the globe. The question then, what volumes of water has the earth already absorbed, becomes an interesting and very important one; one that geologists have scarcely considered.

Dana estimated that the oceans would be 400 feet deeper if these imbibed waters were returned to them. Many of the Atlantic and Gulf States would be entirely submerged. New York city, Philadelphia and New Orleans, would be more than 200 feet under water. Three-fourths of the islands of the earth would be submerged, and a vast portion of Europe and Asia would be one great sweep of waters. Then if Dana's estimate be a true one, and if it be true that all the oceans fell to the earth in pre-laurentian times, the general depth of the same is less to-day by 400 feet than it was then. But Dana's estimate is a very cautious and moderate one; and based upon a thickness of super-crust of about five miles, while modern researches have demonstrated it to be vastly thicker,—some eminent physicists reckoning it as high as 1,000 miles, so that if we reckon it to be 100 miles thick and admit the rocks to contain one-half of one per cent. of moisture, or one-fifth what Dana claimed, the oceans the world over would rise 2,500 feet upon the shores of the continents; or admitting his estimate of rock moisture to be correct, it would be increased to 8,000 feet.

Hence it must be seen that if all the terrestrial waters rolled as a primitive ocean around the earth, they have diminished from 2,500 to 8,000 feet in depth during geologic times. How many of my readers are ready to admit that the primitive ocean was, on an average, deeper than it now is? Whatever testimony we can gather from the rocky volume, points to the conclusion that that ocean was very shallow. Hence we are irresistibly forced to believe that the oceans fell to the earth in great installments. If the seas of the earth were never any deeper than now, then 2,500 feet (or the 400 feet of Dana's reckoning) fell after the solid crust had imbibed that much, and I am inclined to think the intelligent reader will yet agree with me, that the ocean is now many thousand feet deeper than it was even in devonian times, to say nothing of the cambrian age.

But whatever be the present volume of the subterranean waters and rock-imbibed moisture, it is evident that there was a time when the earth did not contain them. So that the present waters on the surface of the globe, do not by any means represent the great expanse of vapors that once enveloped it. Let us remember this conclusion, to which we are necessarily impelled.

But all these revolving vapors fell to the earth, for they are on the earth and within it! and further, every pound that fell, by force of impact and actual mechanical pressure, was registered in the rocky frame of the earth as so much potential energy! This fact has never been regarded at all by men of science. In another chapter it will be seen that this energy was conserved in mountain-making and plication of strata, and that if the materials of the annular system had not reached the earth at different geological ages, there would have been no general upheaval after the close of the laurentian era;—that the mountain systems and the continents of the earth are a measure of the stupendons force expended in the fall of the oceans to their present level, and that the starts and mountain-making ages resulted from successive installments of matter from the *Earth's Ring System.*

When the waters were held suspended in the primitive atmosphere, they would move in that direction towards which the impelling forces would drive them. The centrifugal force of the rotating mass superadded to the driving force of heat, was greater in the dircetion of the equator, hence they would accumulate in the equatorial regions. Let the reader understand how this must have occurred. There being no centrifugal force at the poles, the vapors were kept from falling there by heat alone; but all other parts of the surface possessed some centrifugal energy tending to carry the suspended matter away from the earth in lines, at right angles to the earth's polar axis. This force, to use the language of mathematicians, "decreases the weight of the sea, which is thereby rendered susceptible of being supported at a higher level than at the poles, where no such counteracting force exists."*

Now during the igneous era, gravital force could in no sense operate to bring matter back to the earth, which had undergone vaporization. Hence centrifugal force carried it much farther from the earth and into the equatorial heavens at that time than it could today; for it is evident that if the earth did not rotate at all during that era, the vapors would occupy great heights of the ancient atmosphere, and the centrifugal force then added by actual rotation would cause them to occupy greater heights. That is, there were two forces operating to keep the vapors away from the earth: heat repelling them, and centrifugal force gathering them, as it were, over the equator, and whirling them into rotation with the mass, so that in course of time the vapors must have occupied a vast space in the equatorial telluric heavens, or firmament.

Again it is apparent, that whenever the earth cooled down, so as to allow the vapors to descend, all that did not possess sufficient centrifugal energy to hold them on high, would descend to the earth. This was particularly the case with all polar vapors, if any were there which had not been previously drawn into the equatorial ring. So that the polar heavens became clear of primitive vapors and all other matter associated there-

* Robinson's "Astronomy," page 69.

with before the equatorial matter could possibly descend upon the earth. Then it is plain that we have arrived at an actual age of the earth, in our investigations, when it rolled through space with an annular appendage over the equator, while the stars and the sun looked down upon its surface as they do to-day, in the circumpolar heavens. (See cut page 174.)

At that age rolled the first-born ocean around the earth. From necessity clouds formed and rains descended, winds swept over the earth, summer and winter, and day and night joined in the round of perpetual change; for solar action under such conditions literally forced such things upon the young planet. When men talk about the "heavy," "damp" and "murky" air of primeval time, they ignore the fact that pure gases are invisible, and all floating particles, as mineral and metallic dust or particles of vapor, which render the air visible or murky and damp, must fall as surely as a flake of snow. If such were not inexorable law, what kind of an atmosphere would surround a planet like the earth?

The same causes operated in primitive times to clear the atmosphere as now promote that end. But laying these considerations all aside as comparatively unimportant, we have the grand feature of that age standing out in bold relief, and that feature is the revolving fund of vapors composing the annular system. We must examine this more minutely, beginning, as before, with known features.

The outer perimeter of the revolving vapors condensed first because that part of the mass was farthest removed from the heated earth. But in condensing the vapors occupied less space than before. The mass revolved eastward, just as the earth rotates. The moon fell back westward as it does to-day, drawing the rings-yielding surface after it, just as it draws a tidal wave westward in the ocean to-day. Now as this tidal wave rolls with immeasurable force against the western shores of the ocean, it checks to an exceeding small extent the radial motion of the earth. It produced the same effect upon the rings. It put a brake, as it were, upon the condensing rim of the great revolving wheel, affecting the outer surface more than the interior. But just in proportion as the condensed vapors were checked, would they decline inwards, causing the segregating particles to form a more dense rim or boundary. For their motion being reduced, their centrifugal force would be diminished, and the gravital force comparatively increased, and the rings would begin their decline under lunar influence, and never cease to approach the earth till the last remnant of annular matter reached its surface.

While this condensation went on in the rim, causing a contraction of the same upon the interior mass, particles next to the rim on the inner side would be attracted to it. Thus the condensed rim, for a certain distance inward, would gather matter, as it were, from below, thus forming a hiatus or division between the rim condensed and the great mass uncondensed. Here, then, we have the outermost ring of the system. It can be readily understood how that it would be impossible for this ring of vapor to fall in and unite with the mass. Every particle of the former having a velocity of, say, 24,000 miles per hour, could not unite with the rim of the remaining inner mass, revolving 23,000 miles per hour, any more than two aerolites in adjacent paths, one moving 1,000 miles per hour more rapidly than the other. And never until the first ring had lost so much of its motion as to move with the same velocity as the mass within, could the two bodies become united again.

The same process continued downwards or inwards would form the second ring; likewise the third, each separated from its neighbors with certain definite spaces or divisions.

Can any one following the index finger of philosophy point out any valid objection to this mode of annular formation? When we shall have gone over the geological record we will see that the annular system was very complex. The number of rings formed was great, each separated from the rest as we have shown above.

Here now we have formed the earth's annular system, but let it be understood that we are not now going to build a house upon this foundation—we are only expecting to bring a mass of evidence to prove that this is the foundation! We will continually add testimony as we go along, and we will not build the house until the reader can see an immovable base to build upon. We have demonstrated, so far as a physical question can be settled by law, that there were waters revolving about the earth! We have shown, as I think, the only reasonable mode of annular formation and division.

Let us turn our glass to the skies. Yonder is a bright gleaming orb, nearly 1,000 times as bulky as the earth. Around it revolves an annular system. Across the vast abyss of nearly one thousand millions of miles, we see Saturn's annular appendage, divided into three grand divisions, and these divisions each further divided into a system of smaller ones. The number is not known. The clearest and best telescopes exhibit the greatest number of divisions, so that it is likely a telescope of greater power and clearness than has yet been directed to it, would reveal many more. So here is a world surrounded by a complex system of rings, just as reason teaches the earth was at the close of archæan time. It is divided as philosophic reasoning proclaims. The exterior ring is about 173,500 miles in diameter, and is itself 10,000 miles broad, and the innermost one is more than ten thousand miles from the surface of the planet. Here, then, we have strong reasons for claiming that the process of annular development on both Saturn and the earth was the same. Igneous action was no doubt the only competent cause. We return to the consideration of our own orb, strengthened in the belief that our reasoning is correct.

The earth's annular system has fallen, and we will now philosophize upon the manner of its declension. I have said that the moon put a brake, as it were, upon this appendage, just as it now does upon the earth, and its effect upon its motion extended throughout the system, from the exterior to the innermost ring, so that when condensation and segregation had completed the system, it must have declined bodily toward the planet, and of course the innermost ring reached the outskirts of the earth's atmosphere first. But what would be the immediate effect of the entrance of such a body into the upper regions of the air? Slowly it descended, but the moment it touched and began mingling with the air, its down-progress would be checked. For, however rare the atmosphere at that elevation, it was matter occupying space, and no other matter, however dense, could displace it without encountering some resistance.

This resistance or checking force operating upon the vapors in front while they pushed on from above would cause them to spread into the form of a belt, and this belt would widen and spread from the equator toward the poles. When this innermost ring had so far declined as to be freed from the system, it of course continued to revolve for some time more rapidly than the atmosphere rotated with the earth. Moving in a greatly attenuated atmosphere with an independent motion, there would be two forces resisting its fall, viz: Its own independent revolving energy, and the resistance afforded by the atmosphere, and this latter increasing in a direction toward the earth on account of greater density. Under these conditions the newly-formed belt would float away from the equator in two divisions, one toward each pole, and must have reached the earth's surface in regions beyond the tropics, perhaps beyond the temperate regions and in the polar zones.

Now, it is a well-known physical fact, that the gravital force is stronger in the polar regions than elsewhere upon the earth, from two causes—i.e., the greater attraction, and the absence of centrifugal force at the poles. Thus we see, as in the ring formation the vapors followed the direction of the greatest driving force toward the equator, so in ring declension they returned along the same line and fell where there was the least resistance. Now the resistance occasioned by centrifugal force is zero at the poles, and gravity is greater on this account alone by its $\frac{1}{289}$ part, while at the same time the polar world attracts a body about $\frac{1}{590}$ more than the equatorial, so that the two forces combined make the gravital tendency $\frac{1}{194}$ greater at the poles than at the equator. Hence it is evident that a belt would fall from the equatorial heavens down to the polar world. This will be abundantly proven as we proceed. Now we see there must be a division between belts and space of time between falls of matter from the annular system.

There is another point in belt de'clension we must now consider. When a belt entered the atmosphere, the resistance of the latter would put a brake upon it on the inner side and continue to check its motion until it reached the earth's surface, or wound up its spiral orbit at each of the poles. Hence an equatorial belt necessarily revolves more rapidly than a polar one, and the motion of the polar one more nearly represents the time of the planet's rotation than the equatorial; and further, the slower motion of a polar belt shows that it has been under the resisting influence of an atmosphere longer than the equatorial. In other words it shows that it fell from an annular system over the planet's equator, and has floated away with an actual falling motion towards the poles. I state this as a demonstrahle fact.

Let us return to view the planet Saturn, and see how many of these conditions obtain on that ringed world. We first notice that it possesses a number of belts, pretty well defined. We see they are separated by visible divisions or partitions which necessitates the conclusion that they were separated in the annular system, for we know of no force to separate them after they once left the annular and assumed the belt form, until they fall at the poles. We also find that the polar belts move more slowly than the equatorial! Here, then, are two important links of evidence pointing directly to the conclusion that these belts of Saturn have descended from the annular form,—that they are revolving in the outskirts of Saturn's atmosphere.

In order that the reader may more fully understand the importance of these witnesses, speaking from the heavens and bearing emphatic evidence from analogy that annular development is a philosophic and necessary part of planetary evolution, we must more minutely examine the only annular system now visible, and also bring in the invaluable testimony of *J*upiter, the "King of planets" and giant of the solar system. We must be allowed the privilege of drawing conclusions respecting the former condition of the terraqueous globe, from present known condition of her sister planets. I believe the birth, growth and development of worlds are regulated by inexorable law, and if one planet was ever surrounded by rings, a sister planet under the same circumstances, ruled by the same dynamic and static conditions of force, in process of development, must also be attended by rings during some stage of its career. Not that I ignore the fact that circumstances varying must vary the resulting phenomena of ruling forces, but the great principles of planetary growth must obtain on all planets. It is, for instance, as essential that ring-formation should follow igneous action, as the oblatoidal form of a planet should follow its rapid rotation. They are pure results of acting forces everywhere apparent in the solar system, from the great burning, seething and smoking sun, to the utmost and smallest satellites. If we can detect this universal disposition in the worlds around us, we may rest assured that our own has passed through the same grand cycles of change. Nay, we may in fact read the geological history of the earth in the ringed and belted worlds of the solar system.

It must now be clear that these features exhibited by the belted vapors of Saturn and Jupiter, are vital considerations. Modern science has established beyond a doubt the fact that the motion of their polar belts is slower than the equatorial. From this we are forced to the conclusion that they revolve nearer their primaries.

If those belts could by any possibility increase their motion they would rise and revolve in a larger orbit. That is, they would move from the poles toward the equator. On the other hand, if the equatorial belts should lose the smallest part of their motion they would sink along the lines of least resistance and greatest attraction—i.e., toward the poles. Now can it be possible in a universe of unchanging law, that one planet could become the possessor of a ring-system unless the causes that formed it were universal? Can it be possible that the earth, under the influences of these universal causes, has not passed through the same mode of planetary evolution? I can no more doubt the universality of this process, than I can doubt that an apple would fall from a Saturnian or Jovian tree; and when we see, that in addition to this necessarily universal annular development, the condition of the primitive earth demands such development, we are not even allowed to entertain a doubt upon the subject. If the laws of gravitation be universal, the causes of annular formation are also, and effects must follow. It may be said unknown conditions may modify the operations of the law. Certainly this is true, but they may also modify the operations of the law of universal gravitation;

yet, where is the man who doubts its universal application in the midst of all modifying tendencies?

From this it must be seen that the mere fact that Jupiter's and Saturn's polar belts move more slowly than the equatorial, is positive proof that they have moved from the equatorial regions, and therefore there is a perpetual tendency in the solar system now for all belts to fall at the poles! Here, then, we are simply impelled to admit that the original form of all revolving planetary belts, was annular, and that they were located in the equatorial regions of all planets during some period of their history. The supposition also that these belts must reach the surface of the planets in stupendous downfalls, during intervals of immeasurable time, receives here an emphatic avowal.

Thus by following the path pointed out, by the unerring voice of law, we may look upon those giant worlds, and read a history of the mighty changes that made our world what it is to-day. For unknown ages rings and belts attended the earth. One by one they declined and reached its surface around the poles. Grand stupendous arches spread over the face of the firmament when no man was here to see; when the wild denizen of a wild world alone roamed its boundless wastes, thoughtless of impending calamity. When we gaze upon the fearful and terrifying elements, when cloud meets cloud, and deep frowns on deep in the battlefields of nature, what puny things we are in the wondrous arena! But suppose we dwelt to-day on a ringed world, and could see all these features and conditions a thousand times intensified! We would stand appalled at the fearful grandeur and majesty of world-making.

We must look at our earth in its spasms and eternal

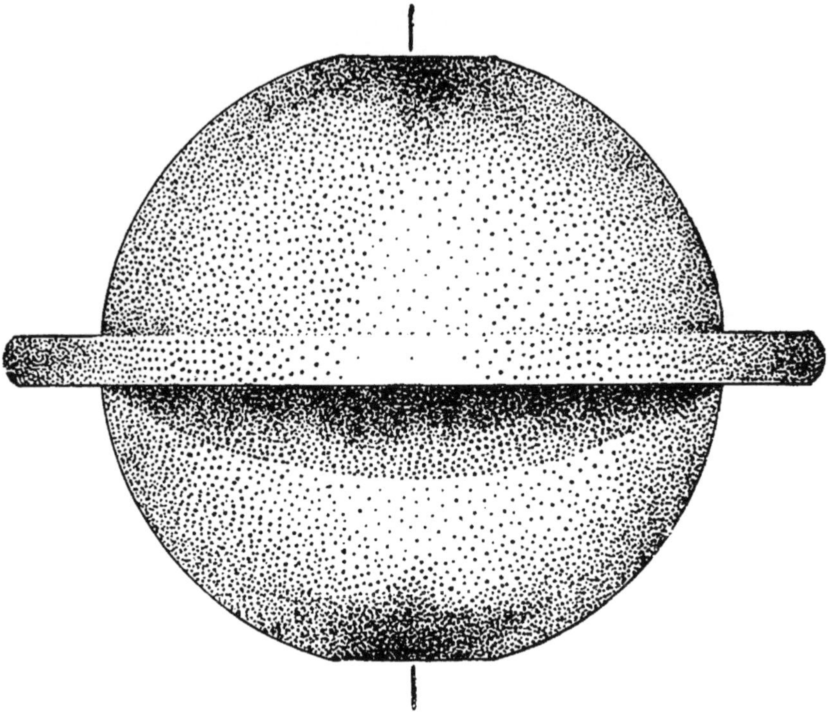

Fig. 2. EARTH COOLED FROM A MOLTEN STATE.

(ITS RING SYSTEM FORMED.)

After the lapse of immeasurable time, the earth had cooled down, forming a firm foundation for subsequent deposits. The great mass of expelled vapors had condensed. Some of these had returned to the earth's surface, forming the first ocean,— a world expanse of waters,— and a world casement of sedimentary beds. In that ocean the first forms of life appeared. High over the equator, as if anchored to the skies, a vast ring system had formed from the higher and lighter elements, which gravitated each to its proper place in the system, according to its specific gravity. Fig. 2 represents this ringed world, with its rings turned edgewise to the observer, and the planet covered with a universal ocean, that ocean teeming with rudimental life, and the sun shining on the earth much as it does to-day.

revolutions if we would embrace half the meaning of annular work in by-gone ages. While rivers flow and bear their burdens to the sea, while the all-devouring waves prey upon the continents and are unceasingly at work in building up strata in the seas and lakes of the earth, we must not forget to acknowledge the tribute of the earth's annular system in building up the sedimentary beds of the planet.

CHAPTER III.

SOME CONSIDERATIONS RESPECTING SATURN, JUPITER AND MARS, AND THE EVIDENCE THEY SUPPLY IN SUPPORT OF THE ANNULAR THEORY.

Saturn is not quite seven hundred times greater in volume than the earth, but he is so light,—having a specific gravity less than threefourths that of water, that he is only about ninety times as heavy. Proctor says: "Gravity at his equator is almost exactly equal to gravity at the earth's surface. Near the poles there is a marked increase in the action of Saturnian gravity, insomuch that a body weighing ten pounds at his equator would weigh about twelve pounds at either pole." It is more than likely that Proctor was mistaken, as it must be conceded that we have never seen the actual face of Saturn, and therefore do not know how rapidly he revolves. Hence all notions as to the comparative polar and equatorial gravital forces, are necessarily vain. He also speaks of total solar eclipses, in the latitudes corresponding to those of London and Madrid, of five and seven years' duration. But the solar orb is never visible to the Saturnians; for, the over-canopying fund of vapors must exclude it from view. It is, however, easily demonstrated that the inhabitants of Saturn, from this very circumstance, are in the midst of eternal day, from the total diffusion of light throughout his revolving envelope, which from necessity becomes an actual light-bearer. All the sunlight received by Saturn is poured into his belts and rings. Every floating particle of vapor or mineral aids in the total diffusion. And the student in optical science will readily understand, how there could be but a sign of the alternation of day and night, on a planet surrounded by vaporous belts and light-bearing zones of revolving matter.

A planet surrounded by a lofty vaporous atmosphere can have but the merest shadow of night while the solar beams pour into it. Neither could there be alternation of seasons, while the solar heat entered such a revolving envelope. Let us examine this a little farther. This difference of polar and equatorial gravity is true as to the envelope of Saturn and shows emphatically the necessary conclusion that the belts must gravitate to the poles in order to fall. Thus, Saturn's equatorial belt presses directly downward with a certain force; but its own centrifugal force, and the resisting atmosphere prevent its motion in that direction. Now at the same time, while this belt is equipoised, in mid-heaven we may say, another force is actually exerted to pull it down via the poles. A lateral motion must be the inevitable result, and this must end in a universal canopy.

It is also plain that, if all Saturn's belts, except his equatorial one, should fall, the single belt, if large, would, in spreading toward the poles, shut out the direct sunlight from the surface of the planet.

Admit a single beam of sunlight in a chamber of midnight darkness filled with steam, and you will see the whole room illuminated. The steam carries the light into the darkest corner. A jet of water, illuminated just as it leaves the hose by the calcium light in the midst of total darkness, will appear as a beautiful stream of light as far as it can be thrown. Any one can perform these experiments for himself, and prove to his own satisfaction that watery vapors, snowy particles, or almost any floating particles except actual absorbents of light, are actual light-bearers. Then a great mass of attentuated clouds or vapors, unpacked by tempests, high above Saturn's surface and extending all around him, one half illuminated directly by the whole light of the sun, must inevitably carry the light of day around the planet; or even a jet of water projected around a planet, would appear as a light-giving ring, and if that ring were extensive enough, it would annihilate night. Not that the parts of the envelope of vapor, or jet of water farthest from the sun, would be as light or luminous as the rest, but that the columns rising from the eastern and western skies, brilliantly illuminated and spreading out fan-shaped in the zenith, would illuminate the planet's surface. While in the case of a wide belt, or a universal envelope, the light from the eastern sky would mingle more profusely with that from the western, and the illumination would be so general that it would be scarcely possible for day and night to alternate as we see now on earth. While one belt remained in the Saturnian or Jovial heavens there could be no true night there. If a single moon shining on earth can so clearly dispel darkness by its reflected light, that one can some times read a common print at midnight; what must be the luminous effects of a universal light-bearing canopy?—a reflector equal to thousands of moons. Even when our atmosphere contains more floating particles of vapor, or cosmic dust than usual, its daylight is sensibly extended; and we can readily understand then how our atmosphere might become so full of aqueous particles, as to extend morning and evening twilight far into the night. But such particles in an atmosphere no more extensive than that of the earth, could have but small effect compared with arching vapors high in the heavens. If we could transfer a cloud from the full light of day with all the light it contained, into midnight darkness, how brilliant it would appear! How it would illumine the clouds around it! So that in philosophizing upon the conditions of our sister planets, astronomers, I believe, have erred by neglecting these facts. Let the reader note this philosophic deduction of perpetual day, for it will come in as startling evidence in its proper place.

One of the strongest points I have to present to show that the bands and belts of these giants of the solar system move almost independently of the central orbs, is the velocity with which they apparently rotate. It is admitted by most astronomers that we cannot see the actual surfaces or bodies of these planets. Then it must be admitted that we do not know the length of a Jovial or Saturnian day. We do know the length of a day on Mars, or on Venus, even to the fraction of a minute, and we also approximately know the time of the rotation of Mercury, and that these three planets and the earth rotate in about the same length of time; none varying more than a few minutes from a terrestrial day, or nearly twenty-four hours. Now it does not seem likely that Saturn, about 700 times more bulky and ninety times as heavy as the earth, or that Jupiter, more than 1200 times as large, and out-weighing the earth three hundred times, would, in the same system, and under the same laws, rotate more than twice as rapidly as any of the four interior planets named. It seems inharmonious. If, therefore, we should assume the Saturnian day to be about 24 hours long and knowing its exterior envelope to rotate in about 10½ hours, we should find evidence to support this assumption, from the very highest authority, my readers will certainly allow me the liberty of ignoring the long cherished idea that Saturn's light belts are his atmospheric clouds, as commonly understood, and his dark ones but rifts in the same, revealing the body of the planet, as some suppose. If the earth should rotate so that a particle of matter on its equator should move about 290 miles per minute, a cloud would be thrown outward to the very limit of our atmosphere, and be impelled to move in an independent orbit about the earth. This is so plain that the merest novice in astronomy must understand it. But Jupiter's clouds move at the rate of nearly 470 miles per minute. Such a velocity would fling a terrestrial cloud thousands of miles beyond the atmosphere, and cause it to move around the earth. But a cloud on Jupiter's surface weighs about two and a half times as much as on earth, so that a simple calculation will show that with the above velocity, Jupiter's clouds, whatever they may be, are to a great extent independent of the planet so far as velocity or rate of radial motion is concerned. That is, if the same static and dynamic forces exist upon Jupiter as on earth, his bands and belts revolve about him. It also follows that there can be no true clouds in Jupiter's atmosphere. We know that with such velocity, no clouds could exist in our atmosphere, even with all the necessary difference of conditions climinated. We also know that a cloud placed in the outskirts of our atmosphere must revolve about the earth, or fall immediately toward its surface, and occupy its proper level. Can Jupiter be an exception? Hence it seems necessarily to follow, that if Jupiter has any clouds, they are raised beyond the region of floating clouds, and hence beyond the region of storms and tempests.

The same conditions maybe predicated of Saturn, said conditions differing in character, as the forces existing differ in degree. In this planet, however, we have more direct and emphatic testimony. So that while Jupiter leads us to the above conclusion, Saturn forces us to the same. It is well known that this planet sometimes presents what is called the "square shouldered aspect;" that is, in some parts of his orbit it is not only flattened in the polar, but also in the equatorial region.* Sometimes this equatorial depression extends over 65° to 80°. This, I conceive, is readily explained by the planet's vapors revolving in ellipses, which they necessarily must do. In certain parts of the planet's orbit, we see the depressed sides of the ellipse, and consequently a flattened equatorial region; and when the major axes are more inclined to our line of vision, we see, as it were, the ends of the ellipses, when Saturn seems excessively flattened at the poles,—midway between these points, the equatorial and polar regions both are depressed.

Thus the annular theory throws light upon one of the most inveterate puzzles; one that has for half a century perplexed astronomers, and defied solution. I offer the suggestion that no other solution can be found. The same appearance in a less striking degree is sometimes seen upon Jupiter. Thus it seems that the two planets under consideration, are respectively surrounded by a vast fund of revolving matter, and that

---

* Proctor's "Other Worlds Than Ours," pages 168, 169.

this matter in its motion follows the same laws that regulate planetary motion everywhere.

Now it is readily seen that this tendency to quadrangular form on Saturn, becomes irrefragable evidence of annular motion among his belts. But recent observations show that the polar belts of both Jupiter and Saturn move more slowly than the equatorial.

But as before shown the simple fact that the polar belts of these planets have a slower motion, affords irrefutable testimony that they at one time were a part and parcel of an equatorial ring system, and that they also have lost some part of their velocity since they entered the planet's atmospheres, and are therefore continually descending toward the poles, and at the poles are continually reaching the planet's surfaces. Let the reader note this fact.

Thus away out yonder, toward the bounds of the solar system, we see two giant worlds undergoing the same stupendous ordeals that in ages gone by, our little earth experienced. The heavens speak as with tongues of fire, and we hear celestial harmonies proclaim eternal law to the utmost bounds of space and time.

But what are those belts, now revealed by the telescope? What kinds of matter constitute those annular and belt systems? Law replies:—" They are composed of the same materials in kind, that now compose the bodies of the planets themselves."

Now I suppose no one will dispute the claim that Jupiter's belts are, aside from the aqueous matter they contain, composed of the very same elements that now compose the super-crust of the earth. Then it must be they contain silicious, calcareous and carbonaceous matter. But if they contain these, they will in time become a part of Jupiter's sedimentary formations, for as I have before shown they must fall. It is impossible, as any one can see, that such matter should not exist in the primeval vapors of every world, and the primeval vapors are the last form of matter that descends upon an evolving planet. And it must be, as I shall show hereafter, that unconsumed carbon occupies a large space among all such revolving vapors.

I ask the simple question, Is there any other escape from this conclusion? Is it not a fact within the comprehension of every one, that if Jupiter's belts contain calcium, iron, or carbon, that that calcium, that iron, and that carbon, will in the coming ages be located as parts and parcels of the sedimentary beds of Jupiter's super-crust? The laws of segregation and gravitation are the same, we may safely assert, on Jupiter as on the earth. Then as carbon is a constituent,—nay, a prominent constituent of worlds, and a fiery burning condition a necessary condition of those worlds at some time of their career, it follows as plainly as the sun follows its course in the heavens, that as all the last descending materials of those worlds must fall upon their surfaces they must and will become a part of the constituents of the aqueous rocks of every orb that ever was enveloped in such vapors. On this eternal rock I stand, and though the cruel, heartless elements now gathering blackness and fury from the realm of error may sweep me from it, this eternal rock will remain.

Now when we have passed over this ground, and surveyed minutely the close and interesting analogies visible on all sides, we seem forced to the conclusion that Jupiter once had equatorial rings the same as Saturn. But when we turn to Mars, we see his polar ice caps, we see his floating clouds, we see his oceans, and how shall we answer this question? Were the Martial seas always upon his surface? One moment's philosophic reflection must bring the response of "no," from the seat of reason. The over-cautious ætiologist may say, "we cannot tell." Nay, but we *can* tell. The Creator of heaven and earth points us to facts that can lead us nowhere else. Mars is an aggregation of matter, gathered and formed into a globe, as countless millions of other globes, and formed under the same physical laws that governed others, and the earth is a God-given key for the mind of man to unlock the whole. Man is no more sure that the earth was once a glowing and burning orb, than he is that Mars was a fire-born and igneous planet. Then his oceans were his swaddling garments, wrapped about him by the genii of the heavens. Though not so extensive as ours, yet there are oceans on his surface, and they must have fallen thither from the heavens around him. Do not understand me to claim that each planet is a representative of all the rest in all particulars. Mars may have an atmosphere varying from that of all the rest. His seas differ in their constituent salts, etc. His strata may be different in many respects. Nay, even the color of his landscape may be different from ours. These are things that vary under varying circumstances. But there are planetary conditions that must obtain in every planet. We are forced to the conclusion that Mars has an atmosphere! We would conclude thus if we had never detected it. We would be impelled to the conclusion that he possessed oceans if we had never seen them, just as we are impelled to believe that he is under the pale of the law of gravitation; for we can no more ignore one than the other. Men may say that a planet's oceans may be absorbed by its beds of rock, but one might as well deny that a planet has rock, as to deny that it has water. Our moon has imbibed her oceans; the earth is doing the same thing; but the fact that oceans are thus absorbed has no force at all against the claim that all planets and their moons must have water in some form about them, upon them or within them; and that it fell to their surfaces from annular systems.

We see the same process now on a grand and measureless scale in the solar orb. His aqueous vapors must be driven millions of miles from his surface. His heat is so great at the distance of the planet Mercury, nearly 37,000,000 of miles from his center, that water could only exist there in a state of vapor. We know very well in his flaming envelope are glowing, heavy minerals and metals, and must conclude that the vapors of lighter minerals, etc., sublimed in the solar surface, must occupy space far beyond and above the photosphere, or his atmosphere. The spectroscope leaves this beyond a doubt. We all know that analogy requires that there should be above all other elements, a great fund of carbon, surrounding the solar sphere. The Titanic furnace that vivifies the solar system, does not reveal more than a trace of it in the spectrum. Then it must be so high above the sun's surface as not to be detected. We know not how far this carbon fund extends. But we do know as comets approach the solar orb, they grow brighter, until sometimes they burst into actual flames; when the spectroscope reveals the fact that the flames are partly burning carbon. But, I think we may safely say, that law demands that in the solar heavens must be a vast fund of allotropic carbon distilled in the solar alembic, and driven from this fiery center to float as infinitesimal particles in the comet's path; and, finally, when the inveterate fires of the sun shall have died out, as they must in time, these forms of carbon, with associated aqueous and mineral matters, will form into an annular system around that great orb. Its aqueous vapors, or light-bearing bands, will then form bright portions of the system, and the carbons the dark and dusky belts. The sun must be a forming world. What other conclusion can the inexorable and universal laws of planetary evolution lead us to? This is simply the declaration of Deity, inscribed in letters of flame all over the universe. And with this, is written the glowing command, "Philosopher, read these lines!"

Now one retrospective glance. Look at the dark and dusky bands or belts of Jupiter and Saturn. When we know that these worlds have passed through the forge of Vulcan; when we know that those bodies from their inmost depths have been boiling, seething, and tossing masses of liquid fire; and that to their inmost depths, carbon was one of their prominent constituents, and was therefore one of the sublimed and distilled products incorporated with aqueous vapors in an upper ocean, how can we avoid the conclusion that those dark belts are necessarily carbon? Can we by ransacking the great laboratory of nature, find any other element that can be made to take its place? The simple fact stands out prominently to the philosopher's gaze, that so surely as Jupiter and Saturn have passed through a state of igneous fusion, so surely are they now enveloped by bands of primitive carbon, in all its allotropic forms.

We can no more ignore this fact, than we can ignore the fact that carbon is distilled in the smelter's furnace in reducing his ores; no more than we can ignore the fact that the same products are formed in the retorts of the gas-furnace.

A burning world must be a smoking world; and from its ten thousand furnaces must rise vast volumes of unconsumed carbon to mingle with suspended vapors. If we deny this we are forced to admit that the burning or igneous world was enveloped by an ocean of oxygen, which runs counter to law. Hence it is evident that every igneous world—i.e., every annular system, has, or must have had a fund of unconsumed carbon as one of its prominent components. Let the reader remember this, for upon it depends the solution of a momentous problem. As I run over this fascinating line of thought, I am tempted to enlarge upon numerous questions that naturally press into my view. A hundred lateral avenues open up, inviting to enter and behold, but even a brief consideration of them would swell this volume beyond proper limits.* I will, therefore, stop short with the consideration of one of these collateral questions.

Jupiter and Saturn have moons revolving about them. They are located according to unchanging law, at a distance from their primaries measured by their velocities and gravital force. These two forces must be equal, to keep the satellites in undeviating orbits. Now it is an indisputable fact, as every mathematician or astronomer will admit, that Saturn's annular system does to some extent influence the motions of his moons,

* In the "Vast Abyss," or second book in the annular series, these fascinating fields will be reviewed.

and thus aid in defining the shape of their orbits and regulating their distance from the planet. To illustrate: A planet attracts a moon, say, with a force equal to A, and the latter takes up its orbit or path in harmony with that amount of attraction. If, by some process, a ring of any kind of matter should become interposed between the moon and its primary, the central attraction would be increased, say $=$ to B, and the moon would immediately begin to sink nearer to the primary, agreeably to the force $A + B$ exerted upon it. If this ring of matter arose from the planet itself, it would not in the least vitiate the conclusion that the moon was attracted with a greater force than before; for, it would only be a transfer of attracting matter from the primary to a point where it could exert a greater force upon the moon. If the distance of the moon were 100,000 miles from the matter on the planet, before the ring had formed, and the ring were then placed within 50,000 miles of the satellite or one-half the former distance, it would attract it with four times the force it formerly did. And the moon, as before stated, would take up a position a little nearer the planet. Hence the conclusion is inevitable that Saturn's moons revolve a little nearer the planet because he has an annular system, than they would if he had none!

What, then, must be the result when Saturn's glorious appendage declines to his surface? Simply the cord of attraction will be weakened, and the moons will not be controlled by the same force, and they will retire, and after a lapse of ages they will move in orbits farther from the planet.

Hence the conclusion that a planet's satellite must move away from the primary after its annular system sinks is inevitable. Now the point gained by this discussion is becoming apparent. Eminent astronomers claim that our moon is moving away from the earth with a motion very slow, but "exceeding sure." If this claim be a valid one we must conclude that the earth once had an annular system which fell and allowed the moon to recede. Here, then, we have very important testimony bearing upon this point. For, if the moon is receding, unless it can be shown that some other force could produce this recession, it becomes proof of itself, that the earth had such an appendage.

I am aware that astronomers are to-day making the claim that this recession is caused by a reaction of the tidal wave upon the moon. (The reader must not conclude that this recession and consequent retardation affects the periodical acceleration, and retardation of the moon is caused by a change in the eccentricity of the earth's orbit.) That the moon is retarded first by a check which the progressing or rather swinging wave exerts upon it, then she moves away in response to the demands of diminished motion. In other words, the erudite conclusion is, that if a satellite be checked in its motion it must move away from the primary. Then the slower the moon's motion about the earth, the farther off it must move, and consequently the greater its velocity, the less the orbit, which is simply not the case.

Now it is very true that if the moon recede from the earth with its present velocity unchanged, it will move in a greater orbit, and of course consume more time in a revolution; in other words, its motion will be apparently retarded. But a satellite cannot recede because its velocity is retarded. Astronomers have misapprehended the nature of the problem. In order to show that I do not misrepresent, I will quote from R. A. Proctor.* "Delaunay pointed to the tides as a probable and sufficient cause of this change,—the great tidal wave carried, not bodily, but still swayingly against the direction of rotation, checking the earth's rotation spin slowly but exceeding surely. Next it *was shown* that, accompanying this change there must be a gradual *loss of lunar motion, accompanied by a gradual recession of the moon.*" (Italics mine.)

If this be true, law is law no longer. If a moon recedes because it is checked in its motion to a slight degree, its recession will be greater if the retardation be greater, and if the retardation be sufficient to stop its motion entirely, its recession will be in a tangent to its original orbit. We can make nothing else of this astronomical claim; for, let us remember that the tidal wave first retards the moon's velocity,—causes a "loss of lunar motion," and then it recedes! Either this or the attraction of the tidal wave causes it to recede. With all due regards for the noble minds which have been puzzled over this problem, I must say astronomers are mistaken.

If a satellite's motion be retarded it will decline toward the central body, and the greater the retardation the greater the decline, until it falls to the primary. But if a satellite recede by loss of attraction from the central body, it takes longer to perform a revolution. Hence we say it is retarded, though it move as rapidly as ever. If it moves inward on account of an increase of attraction, it revolves in a smaller orbit,

* "Eclectic Magazine," May, 1882, taken from "Contemporary Review."

and consequently in less time, and we say its motion is accelerated, though it move no faster than before; for it is plain that if it should move faster it would have increased centrifugal force, and go off into a greater orbit.

If astronomers are correct in this view, then the moon must in time leave the earth entirely. If they are correct, there was a time when it was so near the earth that its oceanic tides were reared mountain high and made to sweep over the continents twice in twenty-four hours. But I beg leave to say that this does not seem, by any means, to be the declaration of the geologic record. The learned Dr. Newberry has shown how erroneous this conclusion is, by a thorough and complete survey of the geologic past.

What cause, then, must we assign for the moon's recession? I cannot call it "retardation." It can recede only by a decrease of attraction from within; for an outward attraction would only cause a local perturbation susceptible of self-correction. A comet might cross its path, and for a moment exert itself to check, or increase its motion, but these ephemeral visitors are harmless as a puff of wind, and if they should check the moon's velocity, it would decline, not recede; and if they should increase its motion, it would recede, without showing retardation. Where, then, can we find a competent cause for the recession of our satellite? If it be not in the fall of the earth's annular system, then, I presume, it can never be found; and concerning this problem, the universe will be as voiceless as death. The conclusion then is simply overwhelming that the earth once had an annular system. Let us now give our attention to this.

CHAPTER IV.

THE GEOLOGIC RECORD EXAMINED.

Having shown that the prevailing idea maintained by geologists in all their reckoning and conclusions,—i.e., the idea that the ocean of water now on the earth fell in its entirety before the aqueous crust was formed, is necessarily erroneous; in short, having established the annular theory upon a foundation peculiarly strong, it certainly demands a respectful consideration at the hands of thinking men, even if no further proof existed.

To go over the geologic record regularly, and point out all the important features, directing to the order and condition of things here hypothecated, would fill a large volume of itself. I will, therefore, take up some of the most important ones,—those also most familiar to and most likely to be comprehended by the ordinary reader.

I have made the claim that the earth's annular system was necessarily a complex one. If the igneous earth had been hot enough to vaporize and suspend water only, then it is plain that the great primeval atmosphere would have contained those vapors only. But those vapors themselves must have contained dissolved silex and quartz, from the fact that hot water and hot vapors will dissolve it. But as we well know the heat of the primitive earth was immensely greater. Then it is certain that its atmosphere must have contained whatever else was vaporized and suspended therein; and thus under law, when the atmosphere became cool, it deposited upon the earth what it contained in the heated condition. So that when I advance the claim that much of the sedimentary beds built upon the laurentian and older rocks were simply precipitates from the annular system, all must see that it simply is impossible that such should not be the case. So surely as hot vapors can contain more mineral matter than cold, so surely did the cooled vapors of the primeval atmosphere deposit vast quantities of mineral matter on the earth when they fell to its surface.

Now all the fusible and vaporizable minerals in the earth's crust must have existed to some extent in the upper vapors; just as all the minerals and metals in the sun must be represented in the heated vapors around it. And every other hot and burning world must exhibit the same thing. Let any man reflect, but for a moment, and he must admit that the present state of physical science demands an unqualified assent to this claim.

There were calcium and oxygen and carbon in the primitive atmosphere. Then there was carbonate of lime; and these elements existing in measureless abundance, necessitates a vast amount of the carbonate in the system. There were iron and sulphur. Consequently these also existed in the upper ocean, as metallic and mineral salts, and it simply seems impossible to avoid the conclusion that the annular system was a vast ocean of homogeneous and heterogeneous matter. By a more laborious and critical examination this conclusion would assume the phase of a positive demonstration, but I need not burden the reader with it now. A certain degree of heat in the burning earth kept the aqueous vapors suspended on high; a greater degree of heat sent up, in their order of fusibility, the minerals and metals of the earth, as they bubbled up as vapor from a boiling crucible. For the present, admit this conclusion, and as we proceed the necessity of it will be apparent.

How would this vaporized and suspended matter arrange itself, as the earth cooled down and the mass contracted? Obviously the heaviest and densest matter—the heaviest minerals and metals—would locate more largely in the innermost part of the system, or nearest the earth. Doubtless all kinds of matter, even metals to some extent, must have remained disseminated throughout the system; but bulk for bulk, the inner part must have been the heaviest, because laden with the more refractory metals, etc. For instance, the innermost ring must have contained more iron than any other ring of the same bulk, while at the same time iron particles of a different state of purity, because of certain combination, and consequent of varying gravity, must have existed in all the rings, except, perhaps, the outermost one, which must have been nearly free from metals, but yet must have contained distilled carbon particles of the lightest form as soot sent up from the smoking earth. Hence when I make the claim that the deposited minerals and metals in the earth's crust follow an order of arrangement which was chiefly determined by annular arrangement in the nebulous atmosphere, the reader must see what kind of order it must be!

We will suppose that the innermost ring of aqueous vapors and their associated matter, comprising all matter within the limits of 20,000 miles, fell after the earth cooled down. But if it fell, it fell because it had not sufficient revolving momentum to keep it above; while all matter still farther from the earth had more momentum and must have remained longer in the annular form. But this innermost ring, when it reached the earth, and mingled with the terrestrial seas, produced an augmentation of the oceans already thereon, and the iron and other heavy metal contained therein, must have formed beds at the bottom of the seas, as an actual precipitate or sediment, and consequently much of the earliest sedimentary rocks must, if our theory be true, contain the heaviest minerals and metals of the crust, and also in the purest form! Is it necessary for me to bring forth evidence to prove that this is the actual state of affairs? Every geologist knows full well that this is the case. He knows that the archæan beds —the oldest formed that have met the gaze of science —are above all others, eminently metalliferous; and that in those beds the metals are in the purest state! Why? Iron Mountain and Pilot's Knob, the grandest accumulations of iron upon any continent, are beds of nearly pure iron, planted amid laurentian piles. In this foundation lie the heaviest masses of lead and galena ore. The copper, iron and other great deposits of Lake Superior and Canada are in the same old beds. In short, wherever these old beds are found there you will also find the metallic beds. And, moreover, they are aqueous, or sedimentary beds! Dana says, "These rocks are universal." They are the metallic sills of the earth. They form a mighty casement, or metal band around the world. Can this fact be philosophically explained without the aid of the annular theory? This iron and this copper and silver and gold, etc., were distilled in the fiery furnace of the primitive earth, and sent up amid the aqueous vapors on high. What else could have planted them in strata of aqueous beds? What else could have made them so nearly pure? What else could have made metallic beds in the outside of the earth? Let him answer who can.

Since we have found amid the old sedimentary beds the very products of primitive distillation which our theory demands, and found, also, in the very condition it requires, and since we can find no other source at all competent, we are simply forced to conclude that the first ocean that fell to the earth must have been strongly impregnated with iron and other heavy metals. They are precipitates from water. But how was water impregnated with them except through the aid of inveterate heat and the annular system? The present distillation of iron by the aid of vegetation proves only one thing, viz: that if such a puny combustion can dispel it, the primitive fires of the earth must have done immensely more. The processes are the same, differing only in degree—the work of combustion, one puny and almost powerless, the other stupendous and titanic.

Thus we may look back through the vistas of time to the primitive and immeasurable age of change, when the first ocean rolled its mineral-laden waves around the earth. In course of time it deposited its load upon the earth. It was a casement of immense thickness, requiring an immensity of waters, much of which we must conclude was absorbed into the rocky frame of the earth as its fires retired within.

It is now apparent that just as we enter on the threshold of this investigation we must meet with the very features our theory requires; but there is much more in this geological horizon. In the northern hemisphere the archæan beds are heaviest toward the north. Now if they were thickest and heaviest near the equator, the annular theory would fail to explain it; but a moment's reflection must show that it does explain its northern development, as no other theory can.

Immediately upon the decline of an equatorial ring into the lofty region of the attenuated air, it is at once converted into a belt, and it gravitates toward the poles, the points where gravity is strongest and centrifugal tendency zero. Hence it must follow that but a small part of the annular system fell in the equatorial world, but more largely in the temperate and frigid zones. Now the geological world well knows that the archæan beds are conspicuously heavier in northern lands; and another condition necessitated by our theory is found, just as we want to find it. It is this kind of evidence that will in the end establish my claim upon a rock that nothing can shake. It is plainly evident that if all the primeval vapors fell in archæan time, as geologists claim, then all the matter that impregnated them must have been deposited in a heterogeneous mass, and not in distinct beds as we find them. There simply could not be those grand and stupendous beds so characteristically different from all subsequently formed strata if the conditions were then as in subsequent times; and philosophic geology demands the very conditions I have pointed out, in order to account for the relationship of beds formed in different ages.

But now, in order that the common reader may be able to understand the points here made, let us admit all the upper vapors to have descended before the beginning of paleozoic times, and therefore from one vast and boundless expanse of waters a mighty bed of precipitated materials fell and formed the azoic beds. This is plain enough. And then all subsequently formed beds were torn away from these earliest beds, and placed elsewhere. This is also plain, and is everywhere admitted by geologists. It is a conclusion forced upon us, as an inevitable result of the above assumption. Then the silurian beds came from the pre-existing beds. But is there a geologist who, after having examined the silurian formations of the world at large, and the archæan beds wherever exposed, would say the former are the debris of the latter, unless forced to such a conclusion by his fatal assumption? By what natural, or even miraculous process could the azoic strata give rise to such a casement of silicious beds, as is well known, forms the base of the silurian in almost all lands? By what natural process was crystallized silica torn from among the carbonates of lime and dolomites and metallic strata, and deposited around the earth without depositing the lime, metals and other minerals in the same beds? Now, if the Potsdam sandstone, and its equivalents in other lands, were formed from the ruins of other beds, the ruins don't show it. But we will let this matter drop. Let these sub-silurian beds be the ruins of pre-existing beds, placed as a mighty covering around them, thus sealing them away from the ocean's devouring waves. But now with this covering, how did the silurian waters get their lime? Did the same waters that before robbed the archæan piles of their silica, and disdained to touch their lime, now after those piles were covered up, begin to rob them of their lime and refuse to touch the silica? They either did this or they robbed the silicious beds at the base of the silurian of what they never had, i.e., the stupendous fund of silurian lime. This matter will not be rendered a particle more philosophic by admitting that the great silurian beds were derived from terranes now buried in the depth of the sea. For it is scarcely possible that continents once the highest should sink and become the lowest. But we will let this subject rest too. Let it be admitted that the silurian waters did obtain their lime somehow from the archæan beds. Now let us see how this occurred.

The lime in the archæan strata is more largely magnesia than otherwise, and therefore the first silurian lime must also be magnesian! But it is not! What are we to do? The lime beds nearest the basic beds of the silurian, at least on the American continent, are almost pure carbonate of lime. How did the silurian waters work through its silicious fundamental beds to the dolomites or magnesian lime, and then taking them up deposit them as carbonate of lime? Now, geologists very well know that this is very wrong. But the difficulty is immeasurably increased when we find that after thousands of feet of lower silurian beds were laid down, and among them the heavy carbonates, I say afterwards, high up in the series, we do find an abundance of limestones so highly magnesian in character as to be denominated dolomites. These facts are too plain to be buried. They stand as mountains across our way. The facts are simply these, and no man will deny them: if it were possible for the silurian beds to be the ruins of archæan terranes, they are not laid down in the order demanded by law! The carbonates where the dolomites ought to be, and vice versa. How did the upper lime beds or dolomites get where they are? If they were originally built up among the archæan, and covered up with thousands of feet of carbonates and silicious beds, how did they ever get out? And why did they not get out when they might have done so—i.e., before other beds locked them down forever?

Here, again, the annular theory gives a felicitous explanation. The waters from which the silicious beds were deposited contained this silicious matter as a mineral distillation, before they fell to the earth; and the waters from which the carbonate of lime was deposited contained that lime when they were on high. The ocean from which it was precipitated was strongly impregnated with carbonate of lime, and must have obtained that lime when the vapors were hot. But the ocean which built up the magnesian lime-beds of the silurian was a different ocean, and made so by additional waters from the annular system. This is abundantly attested by the extermination of species, which always shows a new environment.

When every intelligent man must know that if the earth was in an igneous condition, the matter composing these beds, or at least such matter, must have been expelled from the telluric furnace, and that such matter—matter that had never been formed into continental beds—must have settled somewhere in the ancient ocean, it is the merest folly to claim that all the matter of the aqueous beds was derived from pre-existing beds by aqueous denudation.

There was, no doubt, in all ages denudation and transfer of native material in the formation of beds, but we must not forget that during all these ages a fall and precipitation of exotic matter—tellurio-cosmic matter—aided in the work! It is easy to understand that if the silurian dolomites had been placed next to the dolomitic beds of the laurentian, the annular theory could have had no support, and would be easily overthrown by the fact. But since they are placed just where philosophic geology demands, and yet where the current theory utterly fails to explain, geologists must yield their claim.

We have here, then, the strongest circumstantial evidence that all through these early ages, the upper vapors were falling to the earth and depositing their contained matter upon it. Thus independently of our mathematical demonstration we so far see that the geological history, in its very dawn, declares the essential facts of the annular theory.

Having then, as I claim, laid the foundation of this view, in such a way that no one will attempt to attack it, who has a particle of regard for law, we will move across the mighty abyss of time that rolls its dark flood between the azoic and the present, and lay another foundation on this side the stream, and then we will erect the super-structure intended to span the mighty void.

It is plain, that if after having shown that the earth had an annular system in the very dawn of the ages, I should also show that after man came upon the earth, some remnants of that system still remained on high, then the whole geologic world was built up largely under its influence; that is, that the earth possessed rings and belts throughout all the geologic ages.

We will briefly sum up the conclusions hitherto deduced from the firmly established and generally admitted fact, that the world once passed through the ordeal of fire, or igneous fusion.

1st. All terrestrial waters were held in suspension during that age of inveterate heat, far removed from the surface of the boiling, flaming and smoking mass of the earth.

2d. This suspended ocean of vapors, rotated as a part and parcel of the earth—a primeval atmosphere of great complexity of materials—in the same time that the earth then rotated, just as our present atmosphere now does.

3d. This suspended matter in the course of time gathered in the earth's equatorial heavens, and upon condensing necessarily contracted and segregated into rings, which revolved independently about the earth, thus causing a great lapse of time between the descent of the first, or primitive, ocean of water nearest the earth, and those waters most remote in the annular system.

4th. The waters remaining on high, after the interior waters or first ocean fell to the earth, fell in a succession of stupendous cataclysms, separated by unknown periods of time.

5th. The first ocean was necessarily impregnated with mineral and metallic salts, or filled with mineral and metallic particles to a far greater extent than any other section or division of waters or exterior vapors, for the simple reason that in the system the heaviest vapors would settle lowest or nearest the earth as it cooled down.

6th. All such changes required a great length of time, and a progressive motion of declining matter from the equator, polar-wise; also the bands and belts of the earth's annular system necessarily presented the same general aspect that Jupiter's and Saturn's do to-day.

7th. A succession of concentric rings necessarily requires a vast lapse of time between the declension of one ring of vapors into the outskirts of the atmosphere, and the fall of the next succeeding one; so that each fall, or each ring, after it reached the attenuated atmosphere, continued to revolve as a belt about the earth with an ever-decreasing velocity as it spread toward the poles and over-canopied the earth.

8th. The smoke or unconsumed carbon that arose from the burning world commingled with the upper vapors, darkened them, and formed inevitably, dark bands or belts among bright vaporous ones, as we now see on some other planets.

9th. After a ring of vapors had fallen into the air, it is likely that it may have over-canopied the globe and finally descended to the earth, leaving the atmosphere clear, before another ring reached the atmosphere in its persistent decline.

10th. The apparent retardation of the moon is but a gradual recession of our satellite, caused by diminished attraction as the annular system declined, and the necessary check put upon the revolving rings necessarily caused them to sink and finally fall to the earth, if no other cause of their fall existed; and further, this retardation proves the former existence of an annular system about the earth.

11th. The archæan metalliferous deposits are so located as to be inexplicable by the old theory of aqueous denudation, but beautifully in accord with the new.

12th. The silurian beds, and particularly the order of their occurrence in the earth, utterly refute the idea that they were derived from pre-existing beds. Hence it is evident that during the silurian age there was an annular system about the earth. In other words, it is evident that all the primeval waters did not fall before the dawn of life on the globe.

I here present a chart of the igneous earth and its surroundings, immediately after the heaviest mineral and metallic vapors—which gathered more largely nearest the earth in the system—had fallen, leaving a space of about 20,000 miles between the rings and the surface of the planet. (This vacant space, marked as 1, we can scarcely make hypothetic, as it must seem to be a uccessity in annular formation.) The light parts of the system represent aqueous vapors, and the dark rings vapors darkened by the presence of unconsumed carbon, that necessarily arose from the burning sphere as smoke. Ring 2 represents the heaviest forms of carbon, as graphite, etc., which, according to law, gathered more largely among the innermost vapors than elsewhere. Ring 3, the silurian vapors heavily charged with calcareous and silicious matter, and from which the silurian beds were almost wholly derived during a vast lapse of time. No. 4, vapors of the devonian, carboniferous and permean seas, heavily charged with carbon, hydro-carbons, etc. No. 5, tertiary and cretaceous vapors, containing the lighter forms of calcareous and carbonaceous matter. No. 6, the vapors of the quarternary, containing the lightest form of carbon, now mixed with the glacial drift of the world and imprisoned in polar ice. No. 7 represents the aqueous vapors of the Edenic period, and the Noachian deluge.

Imagine the innermost section of ring 1 to decline from the system into the atmosphere and gradually spread over the terrestrial heavens, in its effort to reach the poles, remembering that all such movements con-

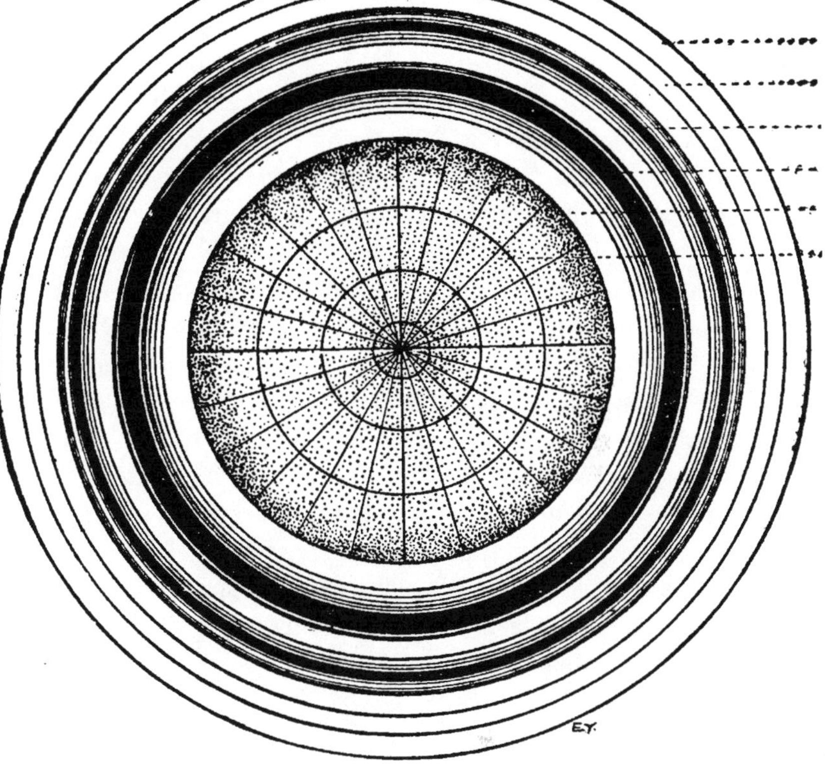

Fig. 3. EARTH AND ITS ANNULAR SYSTEM.

Fig. 3 represents a full-face view of the earth and its anuular system. Here *a* is the earth, *b* the earth's atmosphere, *c* the heavy carbons and their accompanying mineral sublimations, *d* the lighter carbons and hydro-carbons, *e* glacial snows and their accompaniments, *f* outer vapors, principally aqueous and likely in a frozen state. From this outermost ring came the polar snows that chilled the Eden earth, and afterwards caused the deluges of Noah and Deucalion, and still later caused those sporadic incursions of canopy scenes so vividly shining to-day from all ancient scriptures, sacred and profane. In these outermost rings a vast quantity of gold vapors, sent as fire mist to the skies, condensed and forming into nuggets, flakes, flour-gold, and the like, fell in polar lands *with the snows as they fell* and must to-day be found *in and on the very glaciers* that lock down that once semi-tropic region in the grasp of eternal winter.*

The inter-annular spaces of this figure represent similar features in the ring system of the planet Saturn. These are probably filled with invisible air,—an annular atmosphere.

***I have more fully elaborated the annular origin of polar gold in my "Alaska, Land of the Nugget. Why?" and also in "Ophir's Golden Wedge."**

sume a vast length of time, and we may be able to conceive how very often the earth must have been overcanopied as with a greenhouse roof, and how very frequently during the geological ages the earth became a greenhouse world, with intervening periods of flood and desolation. How very frequently the oceanic waters were changed in constitution, and their volume and depth increased. How very much the sedimentary beds were increased in amount by catastrophic additions. But here again let me remind the reader that I do not claim that these additions of exotic matter built the aqueous strata, but that they greatly aided in the work of denudation and transportation of matter, and that hence the time of building being greatly shortened the ages could not have been of so great duration as we have generally supposed.

A critical examination of the aqueous strata will show that they were planned according to the order here represented.

CHAPTER V.

THE EARTH'S ANNULAR SYSTEM AS DEMONSTRATED BY HISTORIC TESTIMONY.

I have intimated that the views I have advanced could not be more strongly supported by the voice of science than they are vindicated by the claims of history. Yet were I to urge biblical evidence to the front, because of my conscientious regard for the sacred writings, it would be assuming a greater authenticity for such testimony than many of my readers are willing to concede. Therefore, in order that it may not be said that I place undue value upon any evidence herein advanced, I will put these writings for the time being on the same level with profane history, however my inclination rebels at the thought. Such evidence, then, as I glean from Genesis, will in this argument be of the same value as it would be if found in the writings of Pliny, Tacitus or Herodotus.

The question now to be considered is: Did any part of the annular matter continue to revolve about the earth until after man came upon it? If I succeed in showing that some of those revolving vapors remained on high, and were perceived by man, then the question will be forever settled, and almost every physical and metaphysical science will have to be reviewed. For it will show that every form and phase of geologic life has so depended thereon as to be modified thereby. It will show that the earth's ring-system, anchored in the terrestrial heavens, when this planet was in its infancy, continued to act the part of a mighty world-carver throughout all geologic time, and lent its titan energies in building the wondrous piles of aqueous beds—the debris of continents and ruin of rings. If the last remnants of the system came down upon the earth in modern times, man would certainly have conveyed the intelligence down to the remotest age, by history and tradition, and the account, if true, would harmonize with law.

Then let us suppose that to-day a fund of annular matter were revolving about the earth. In order to remain in the firmament it would have to revolve more rapidly than the earth rotates upon its axis, and if it were in the outskirts of our atmosphere the resistance of the latter would drag it into belts, and as I have before shown, it would begin an exceedingly slow polarwise motion, in its efforts to reach the earth. It would thus in time over-canopy the earth, forming a universal aqueous roof, becoming a clearly defined and well-known appendage.

Man could not fail to know the nature of that appendage, and seeing the waters already on the earth, and seeing other waters on high, as the source of all waters, he would naturally call the two waters by different names—waters here, on earth, and waters yonder, in the sky; or, waters above and waters below.

Is it not a little remarkable that almost the first announcement made by the Hebrew historian is a positive declaration that "God made the firmament," or aerial expanse, "and divided the waters which were under the firmament from the waters which were above the firmament. And it was so?" (Gen 1: 7.) We are simply given to understand that the writer knew there were two bodies of water—one above the earth and in the sky, and the other under the sky or firmament, or on the earth. No amount of torturing can make this passage mean anything else than the simple fact that a fund of waters revolved about the earth. The merest child knows that no material substance, vaporous or meteoric, could remain in the terrestrial firmament for one moment, unless it revolved about the earth! Science settles this question at once and forever! so that our historian, when he made the declaration that the firmament, or Hebrew atmosphere, became an expanse between two bodies of water, one of which was on high, and the other on the earth, could not have predicated the fundamental truth of the annular theory in more positive terms. Had he said, "We now behold a great deep, or fund of aqueous matter moving rapidly around the earth," he would have said nothing more than he did. The fact that the waters were *above* the firmament demands most positively that they should move rapidly around the earth, with a motion of their own.

How wonderful the thought that the store of sacred history should be opened by the grand conception of a revolving deep! Let the doubter for a moment pause upon this threshold of a new world, and ask: Why is this announcement the very thing demanded by law? He has been schooled in the belief that the earth was once a burning world. Then he sees one of the grand results of that condition; and he must inevitably see that here on the first page of interdiluvian history is shadowed the very fact science has led him to believe; for, if the earth ever passed through the ordeal of fire, there was a time when there were waters above the earth, and waters on the earth. If the historian had here followed the line of thought that an impostor in this twentieth century of the Christian era would do he would have said: In primitive times, directly after the earth cooled down, all the aqueous vapors descended to the earth, and from that day to this no waters have been added to the ocean's volume. Over this the intelligent reader would stand confounded in his attempts to harmonize the different statements. As we proceed it must be plainly seen that the penman would have inextricably involved himself in the plainest stultification. For the sun would have been made visible in the same primitive age, and must have rendered contradictory and false nearly every subsequent statement, as will be seen. On the other hand, the plain, simple announcement of upper waters is in harmony with law, and in harmony with the entire thread of the narrative from beginning to end. Let us see.

Thus we begin our investigation of Genesis, with the announcement, remarkable for strength and simplicity, that some portions of the terrestrial waters did remain on high, until they were recognized as such by man. We see that announcement in utter harmony with philosophic law, and all men must then give it the credit of honesty and truthfulness, though it were the declaration of a Moor, or a Hottentot.

As we proceed we find truthful witnesses clustering around and supporting this great central fact. Right here we learn that "light" came in and garnished the heaven before the sun was seen. (Gen. 1: 3, 4.) This, again, is the demand of law. The upper deep overcanopied the earth, hiding the sun, but revealing his light by the laws of universal diffusion among the vapors. Suppose the writer had said, the "sun now came into view." He would then have contradicted himself on the first page of history. For it is plain that no sun could appear except as a great display of light through the revolving deep. These two statements, then, are co-linked together as important witnesses to the truth of annular formation. Neither of them can be true unless the earth then had an upper fund of waters—a great deep beyond the firmament. How does it happen that this dove-tailing of facts supports the very claim which could be so easily refuted if a single contradictory statement were made?

What was the name of that expanse of waters? The waters on the earth "were called seas." (Gen. 1: 10.) Then it is evident that the "deep" referred to in Gen. 1: 2 was not the waters on the earth, but the waters overhead. This is also evident from the wording of the entire verse. The writer says the earth was void and vacant—ruin and waste—and then turns his attention to the heavens, and says, "And darkness was upon the face of the deep." As all men believed that God dwelt in the sky, or had his throne established upon the upper side of a solid floor, called heaven, we can easily understand why the writer said: "The Spirit of God moved upon the face of the waters, and said: 'Let there be light!'" That light burst in from the heavenly sphere and illumined the upper deep. It would not at all comport with man's idea of the power and nature of Deity, to suppose that his spirit moved on the surface of the "seas" and said, "Let there be light." But if we now take the only philosophic view, viz.: that man knew there was a great "deep" on high, and "seas" on earth, and first described the condition of the earth, then the condition of the sky, and the manner in which the Deity, "brooding on the vast abyss," or deep of heaven, called in the rays of the solar orb, we see again astonishing harmony. The writer of Genesis, seeing the great deep above the firmament, and knowing, from some source, that all the waters on the earth came from that deep, tells us, first, the earth was once "without form and void"; then adds this simple statement that "darkness was upon the face of the deep," when he must have referred to the waters above, which is a declaration in favor of the revolving vapors; and when he again states that the "spirit of God moved upon the face of the waters," while he must have known, or believed, that the Deity moved and lived on high, is another statement, simple and plain, that a fund of waters revolved about the earth. Thus we have two simple announcements in the second verse of the first chapter of Genesis, that the earth had an annular system recognized by man; and again in the fourth and seventh verses it is twice declared, and more positive and emphatic. And again, in verse nine, God said: "Let the waters under the heaven be gathered," etc. Why is the expression *under* used, unless it be to distinguish the "seas" from the "deep"? and another link connects this mysterious history in philosophic and harmonious accord with law. Five times declared, and each time by different means, before half a page is written! It is not a reiteration or tautology, but a simple statement of five different conditions of the surroundings of the ancient world! And each of those conditions predicates an annular fund of waters. I care not whether the historian be an impostor or servant of the Most High, one thing is plain: the very nature of these statements forces conviction upon the philosophic mind. These links of evidence were not penned

Fig. 4. THE FIRST CANOPY SLOWLY SPREADING POLARWARD.

Fig. 4 is an edge view of the Earth's Annular System with its innermost ring having reached the atmosphere in its slow and gradual descent spreading from the equator to the poles. Revolving rapidly around the earth, it is thrown into bands, belts and lines as it forms into a canopy such as the planets Saturn and Jupiter have to-day. I want the reader to note particularly these linear formations and recognize the extremely slow lateral motion toward the poles where all canopies must end their career. An observer standing under this forming canopy would not only see its rapid motion eastward but its slow descent to the poles. Turning north or south he would see a vast arch bent over the vacant air, and he need no longer wonder what the penman of Job 26: 7 meant when he wrote: "*He stretcheth out the north over the empty place*, and hangeth the earth upon nothing."

in order that a man in this twentieth century should prove the truth of a theory; and yet they point with such unerring certainty to this grand and fascinating field of thought that it seems as though the penman had himself reveled therein. But let us see further. If it be true that man thus recognized the upper waters, then he lived in an environment consequent upon that condition, and all the natural phenomena of the inter-diluvian world must have had some relation to the same. If there was a fund of water or vapors above, it must have affected all conditions of life until it fell to the earth. The sun could not be seen as it now appears, until the heavens were cleared of vapors.

In a subsequent part of this volume I will prove that the heavens became cleared at the time of the deluge, and therefore the sun did not and could not have been seen clearly until after that event.

Now let us examine a few more links of evidence gleaned from this fruitful record. Gen. 1: 14 to 19 reveals the fact that the "lesser" and "greater" "lights" made their appearance in the heavens on the fourth day of creation. Laying aside all other consideration one thing stands out boldly to view—i.e., the sun, which physical science declares had existed for measureless ages, did not appear in the terrestrial sky, until after the earth brought "forth grass and the herb yielding fruit." Then it is plain that some intercepting canopy cut off the direct rays of the sun. But, as before stated, no such canopy could exist in such a position unless it had the form and motion of revolving rings or belts. Thus again the plain statement that the "lights" did not appear until the fourth day is a simple declaration that during the first, second and third days, at least, the earth had an annular system! Suppose the narrator had said the sun, moon and stars appeared on the first day. In that case nothing could be more easily done than to prove him an impostor. But the statement that the lights did not appear till later harmonizes with law—with the previously made statement that there were "waters above the firmament," with the demands of the annular theory—an upper deep.

But the reader will notice that he does not state that the sun and moon made their appearance on the fourth day, but simply "lights." The Hebrew word from which the term is derived does not mean sun, nor moon, and evidently refers only to diffused light. The Hebrew words from which these are derived are not used till after the deluge, when the sun was known by man to be both a "lighter" and a "heater." The names, then, sun and moon, not being used, it is evident that they did not even, on the fourth day, appear as they now do, but simply as "lighters," illuminating the vapors. "Let there be lights!" Why did He not say, "sun and moon?" Surely, because the sun and moon were not yet unveiled.

But the writer did use the term "stars," which in almost all ages, according to law, must have shone in upon the earth from the polar heavens. Thus we have mirrored one of the essential features of the annular theory: that the vapors fell largely at the poles. During the prevalence of the upper vapors the polar skies must have been cleared again and again, permitting the stars to shine upon the earth from those quarters. Now a little reflection must convince the reader that the scriptural statement that the "great lights" and "the stars also" appeared on the fourth day, conveys the very idea our theory demands. If the terms sun and moon had been used the statement would have contradicted the statement just made, of upper waters, and would in turn have been contradicted many times in the succeeding narrative. But why this harmony—this unity of evidence? The fact that the term stars is used argues that the term sun would have been used if that luminary could have been seen.

Perhaps the reader now begins to understand why the author was so particular, in a former chapter, in his comments on the motions of the belts of Jupiter and Saturn—i.e., their polarwise decline. Belts could not revolve long in the polar heavens, and would necessarily fall, clearing the circum-polar skies and admitting the stars. Here we see this necessary condition referred to in Genesis. I can conceive of no reason why the name stars should have been used and the names of the two most prominent luminaries entirely overlooked by the historian, unless the stars were seen and the sun and moon were not seen; and as this is the very feature our theory demands with emphasis, the question is most conclusive.

Thus, again, we have to face the fact that the "waters above the firmament" had not yet fallen. The fact, also, that no mention was yet made of their fall argues that they yet remained on high. Thus every step we take leads to a grand confirmation of our views, and in turn substantiates the narrator's account in a way most complete and remarkable.

But the most remarkable and conclusive evidence is yet to be examined. If the waters above still remained on high, and prevented the sun from shining down upon the earth as it now does; if it yet had appeared only as a " lighter," its heat must have been diffused among the upper vapors, and the earth's surface could not have been heated up by its direct rays, but the whole earth under the over-canopying vapors must have been warmed, and its temperature and climate equalized by transmitted and diffused solar heat; just as a greenhouse is warmed by sun's heat transmitted through a painted glass roof. Now this is no vain or idle conclusion; but so surely as the sun's light and heat were diffused among the upper vapors, at the period alluded to, so surely was the earth under a greenhouse covering, and possessed of a climate and temperature harmonizing therewith. The conditions, then, that must have obtained in such a world are substantially these, viz:

1st. There must have been a greenhouse temperature and climate prevailing over the greater part of the earth.

2d. *T*here could not have been storms and tempests as we now have on earth; for the reason that all such phenomena are caused by sun-power—sun-heat falling directly upon the earth's surface. Winds and storm must have been reduced to a minimum; and what is more, rains must have been infrequent, if they could possibly have occurred at all. This certainly is Law.

3d. *T*he solar-beam shorn of its active power, it must have been an age of rest to the earth. There could not have been the alternation of seasons as there now is. Winter and summer would cease to alternate, and there would be one perpetual seed-time, and one perpetual harvest.

4th. Man living in this universal greenhouse would naturally harmonize with his environment, and during that day when solar actinism was shorn of its strength, he must have experienced remarkable longevity; for, it must be remembered that upon solar energy depends every form and phase of life on earth!!

Now let me call attention to a few simple statements in the second chapter of Genesis. The first we notice is the well-known declaration that there was a day—an age of rest. "And God blessed the seventh day and sanctified it, because that in it he had rested from all his work," etc. (Gen. 2: 3.) Can the human mind see any meaning at all in this remarkable announcement, except through the light of the annular theory? Did God, the Creator of all things, ever *rest* from His labors, except those pertaining to the earth and seen by and familiar to man? Did the planets cease to move, the suns cease to burn? Did the solar ray cease its eternal work? Did the clouds cease to move, the rain cease to fall, the seasons cease to alternate, except as looked upon by man shut in from the universe by a stupendous greenhouse roof—by the waters above? All phenomena of nature were looked upon by the infant race as the immediate work of Deity. If the sun could not heat the earth's surface, then that much of the work of Deity was suspended in the estimation of man. Then clouds could not form, and tempests could not rage, and that much of God's work ceased in the eyes of the human race. Then winter could not chill the earth with his icy breath, and the race would see another labor suspended. Fountains and rivers and streams, reduced to minima, would almost cease their labors. In short, if this world was ever enveloped by a fund of vapors "above the firmament," it was characterized by a condition of universal rest! And this is apparently the only possible manner in which the God of heaven and earth could have rested. Man saw these things reduced to this condition as an absolute necessity arising from the presence of the upper deep. And we are again simply compelled to admit the truth of our theory; for, the God of nature—the God of the infant race—could not possibly have rested if the earth had not at that very time a canopy of vapors revolving about it. Now it can be readily seen that if our author had said that God never rested from His labors, his statement, if true, would have overthrown the annular theory at once and forever. How grand the thought, then, that the very condition demanded by it is proclaimed as immaculate philosophy, shaming the mockery and scholastic bigotry of the world! Beautiful, indeed, the concept that the Great Creator presents to the human race, in its greenhouse cradle, a Sabbath typical of that glorious rest prepared for the people of God, where the physical sun will be again shut off, and the "Lord God and the Lamb" will be the light of the spiritual world.

But what is most remarkable and overwhelming is the fact that we scarcely have finished our contemplation of the physical Sabbath, which, above all things, necessitates a windless, stormless and rainless age before we are told that it was a day when the "Lord God had not caused it to rain upon the earth." (Gen. 2: 5.) Such harmonious features must strike the reader with amazement. Every one must see that if there ever was an age in which the earth was not watered by rain it was windless, stormless and winterless! We cannot avoid this conclusion by any human possibility. And, again, we are compelled to admit, however unwilling we might be, that it was an age in which the sun did not and could not shine directly upon the earth—i.e., that the earth's upper waters still revolved about it. At first sight, perhaps, the reader might not think there is much in this. But we must remember that here is a statement of a physical fact, and if we had read it in Ovid only the fact would be the same: that if it be a true statement—if the earth was not watered by rains, but by " a mist "—then the sun's heat was intercepted, and then there was an intercepting body; and since these harmonious statements are all dove-tailed into unimpeachable testimony we are led to believe that this history is the most marvelous ever penned by the hand of man—a history of the earth while yet under the far-reaching influence of the last remnants of its annular appendage. The last ring of vapors in some form had so far declined into the terrestrial atmosphere, as to spread over the earth in its effort to reach the poles; for the last time the sun was again shut out of view. Rains for a short geological period ceased as they had done many a time before; for, it must be remembered, that this was only one of many similar changes through which the earth passed, and which left their records on its rocky frame. The tiny rain-drop has left this testimony upon the living rock. Certain rock-formations say in positive terms that ages before this clouds marshaled their forces in the heavens as they do now, and others are equally emphatic that rains had again ceased, and the earth was a world of verdure unbroken by the reign of winter and storms.

But independently of all these considerations we all know that the warm greenhouse climate of the Eden world is boldly set forth by the writer or writers of

Genesis. There was a warm climate, for man dwelt naked upon the earth. (Gen. 2: 25.) The infant race must have been nurtured and cradled in a greenhouse world. There was a paradise—a garden in which all manner of trees grew, and where all animals named by the Adamite lived. Then that garden was the Edenic or greenhouse world. Now what could have made this greenhouse world? This rainless earth? Just previous to this the world was bound in the icy fetters of the mighty glacier—a sea of universal snow and ice; then it was blooming and lovely, fit abode for the human race in its infancy. It is plain that no feature of the Adamite period is more strongly painted and emphasized than the warm climate of the Eden world. Then another claim of the annular theory is here vindicated. The very climate necessitated by the overarching waters, is positively and emphatically set forth; and we add another link of evidence to the great chain.*

Another thing, set forth in language too plain to be misconstrued is the great longevity of man in antediluvian times. People lived to be 800 or 900 years of age. Now it seems to me I need not tell the philosophic world that if members of the human race attained the age of 800 years, it was primarily because of a modification of solar energy. And as this subject will be fully treated upon in another volume, I will

---

* Here I feel strongly inclined to follow up the Edenic narrative of which there is not a feature that cannot be beautifully explained by the new theory. But it would require a volume to do it justice. If Providence favor, it will be set forth in a future day. Meanwhile, the reader may run over this fascinating field and anticipate the inevitable result—the abandonment of the vain and unphilosophic idea that the Garden of Eden was a local paradise for infant-man.

merely refer to the fact that man's physical environment in antediluvian times, simply impelled long life; and as his longevity diminished immediately after the upper deep fell, and the sun began to pour his beams upon the race, it is evident that his environment changed with that event! In a few generations after the flood man died at the age of 120 or 100, and, finally, at "three score and ten." When we place these facts together, we find in man's great longevity another important link of testimony. Man could not have lived 800 years if his environment was then as it is now. Then it is plain that his environment changed at the time of the flood. But the narrator tells us that the rainbow was then placed in the cloud! (Gen. 9: 12.) Then it is a fact that cannot be disputed, that the sun came into view more clearly at that time, and the environment was changed because it came.

The inference, of course, is that the rainbow was not seen by antediluvian man, which is one of the very things the annular theory claims. The sun could not shine through the annular vapors, or there could never have been an Eden world. It could not shine upon the earth, or it must have rained in Edenic times. It could not shine upon the earth's surface because God had placed the deep "above the firmament," and man lived 800 or 900 years, because of certain solar chromatism and actinism effected by vaporous absorption. The death-dealing properties of the solar beam were sifted out as they entered the revolving vapors. But I cannot too strongly press upon the reader the emphatic and conclusive evidence of the rainbow. If it came into view at the close of the deluge there is no possible escape from the conclusion that the fall of waters cleared the terrestrial heavens of annular vapors! Of this more in another place.

I have said that the antediluvian world was almost free from winds and storms. It was free from them because all such phenomena are children of the sunbeam. Then it is plain that when the heavens were cleared, and the sun shone directly upon the earth's surface, the winds of the earth must then have received their directing impetus. Surely, then, if it had been recorded that winds came into play in the economy of nature immediately after the deluge, contemporaneously with the rainbow, the author of Genesis would have advanced overwhelming evidence in favor of the truth of the theory I advance; and at the same time invest himself in an armor glittering with the priceless gems of *T*ruth; giving value and importance to his history that is accorded to no other ancient book. Will not my readers fully grant this? Would it not have been a glorious summation of the argument in support of the annular theory? But stop! Have we forgotten that at this very time, when the glorious bow was painted on the clouds, or spanning the new-born skies; at the very time the "fountains of the great" celestial "deep were closed," and the "windows of heaven were stopped," "God remembered Noah," said the historian, "and made a wind to pass over the earth." Do we need more and stronger testimony to plant our theory upon a rock that no man can shake? Can evidence be more overwhelming than is found in this grand array of stubborn facts? One glance at the circumstances under which this wind occurred, the first that is spoken of, and perhaps the first that man ever saw, will, must convince the philosopher that it was a remarkable one indeed! The winds of this day herald the rain. They bring on the rain, and the storm dies away in the calm quiet of the equipoised elements. But that was a rain from the fountains of heaven, and when it ceased the sun shone down on the desolated earth! At that moment, all the air-currents began their eternal round. The trade-winds then began their beneficent offices. One-half the earth was then warmed by the sunbeam, that for centuries had no power upon it; and when we consider the stupendous force thus expended, we can no longer wonder that the wind was looked upon by man as the conqueror of the flood. Now the simple fact that it came after the rain makes it a remarkable anomaly, and proves that the flood came from exterior waters.

It can be readily seen that if the wind had occurred as it now does previous to the rain, that it would have forever crushed our theory. The fact, then, that so many harmonious links of evidence join in its support, must give it overwhelming and crushing weight.

But what about the eternal summer of the Edenic world? As the annular theory claims that summer and winter could not alternate as they now do; as the absence of the bow points to the same fact; as a rainless world demands the same; as the Edenic narrative from beginning to end enforces the claim that the earth was characterized by endless summer; there can no longer be a doubt that such a state of things really existed. I presume that a perpetual summer, necessitated by a modification of solar actinism, as it operated only through the upper vapors, necessitated long life; but did the writer of Genesis know this, and did he state that man lived 800 years because he had stated that it had not rained? Did he state it had not rained because the Edenic day was a day of rest? Did he state, "God rested," because he had stated the sun was not seen, when there was light? Did he give the whole narrative, in this grand and inexpressible harmony, with the important declaration, that there were "waters above the firmament"? Now every one must see that all these circumstances, conditions and phenomena, are emphatically necessary results of the presence of upper waters; and that not one of them could naturally have obtained if there were no such watery or vaporous roof on high. And, since eternal law demands, independently of all history or tradition, that the God of nature did place a fund of waters above, how many of us will now put no more confidence in Genesis than in Herodotus? It is as plain as the noon-day sun that the absence of the rainbow in interdiluvian times demands the existence of upper vapors, which the first stroke, almost, of the historian's pen places on high; and that nothing else can explain its appearance at the close of that appalling debacle of overwhelming floods. But tell me, did the author of Genesis designingly state this remarkable truth in order to confirm a dozen previous statements, every one of which is planted on the rock of the annular theory—the waters above the firmament? Every philosopher must know that there is not a particle of truth in this rainbow question, except in this light—and in this light it shines as one of the sublimest truths ever penned.

But I repeat, what about the perpetual summer that this condition of interdiluvian things imperatively demands? this non-existence of perpetual change in the seasons, which the very presence of an over-arching fund of vapors requires? Is there any intimation in this fruitful history that points to a stormless age,—a winterless world? Man dwelling naked in his Eden clime, says in plain language, there was no alternation of summer and winter. His great longevity is unimpeachable evidence in favor of the claim; and the physical sabbath, or day of rest, joins in the harmonious chain of testimony. This eternal summer, it must be seen, is necessary to make the harmony of the historian's account complete.

But it must also be admitted by every intelligent reader, that if such a climate and conditions of seasons existed before the deluge, the fall of waters must have made a sweeping and far-reaching change at once. Eternal spring or summer must have changed in a very short period, to alternating summer and winter, etc. Now if the narrator had even remotely intimated that such a change took place at the close of the deluge, such intimation would certainly be admitted as strong evidence in favor of this theory of the deluge. Especially when coupled with the other new phenomena and changes introduced, as before mentioned, it would be taken from the pages of profane history as evidence peculiarly strong, because of its harmonious union in the great chain of tèstimony. It is then with supreme satisfaction that I turn to Gen. 8: 22, and read in plain, simple terms the very intimation the philosopher would expect and desire to find.

The earth had been desolated for the last time by supra-aerial floods. *T*he survivors of that appalling visitation were introduced to the new environment and ordinances of the skies; when momentous changes were instituted, and new decrees were set forever. Nature's philosophic sign of eternal security, bright and glorious, spanned the new-made firmament. Then the voice of nature proclaimed in the heart and mind of man: "While the earth remaineth, seed-time and harvest, and cold and heat, and summer and winter, and day and night, shall not cease." * It was the voice of God, proclaiming through "nature's vast cathedral," a momentous revolution. Can it be, that the erring voice of tradition even, to say nothing of the unerring voice of law, would have said that summer and winter, etc., should forever alternate after the deluge, if they had been alternating before? Can this announcement be made to mean anything at all, if it point not back to the greenhouse clime of the pre-diluvian period, when the earth was dressed in the verdure of eternal spring; when seed-time and harvest did not alternate, but one perpetual seed-time and one perpetual harvest were the familiar characteristics of the habitable earth?

It would seem that this would be the proper place to refer more largely to these things; and especially to the nightless period of the Edenic world, when both "evening and morning were day,"—i.e., coalesced into one period and called day; but I cannot do this without extending this volume entirely beyond its intended limits.

Now turn one backward glance, and behold the ground on which we have passed. See it thickly strewn with evidence all pointing to the upper waters, predicated upon the first page of Genesis. Note the indisputable fact that all these things proclaim a deluge to come in the ordering of Nature's God. Note the additional fact, that pointing to a deluge, they also

---

* I am inclined to render the Hebrew, "Shall cease no more."

point beyond it, to a radical change after the flood, which change in turn points back to the grand cause of all, the annular waters. But one change that took place because of a fall of waters would stand as strong evidence of that great cause. What, then, shall we say of all the changes that followed the deluge? Why did the bow come then? Why did man's longevity decline at that time? Why did alternating seasons come into play after a deluge? These things must be explained by philosophic law; and I stand under the protecting wing of science to proclaim that philosophic law declares that these things, individually and collectively, demonstrate that the antediluvivan world was over-canopied by the annular waters.

It can now be seen that the very manner in which these statements are made, adds great force to their testimony. They all harmonize and point to one central thought. Not one contradicts another, and the final close is the magnificent triumph of the historian, whose unvarnished statements are each demonstrable by inexorable law.

It must be seen at a glance, that the manner in which light came down, as declared in the third verse of the first chapter of Genesis, predicates the existence of upper waters, so does the "dark" "face of the deep" as before referred to. But upper waters predicate a deluge. Consequently a deluge is indirectly announced and prophesied by both these statements. Then immediately following them comes the positive statement in the 7th verse, that there "were waters above."

I care not who penned these consecutive statements.

They are from their very nature pure truth; rendered doubly pure and refined by the philosophic requisition, that each separately, and all combined, declare that a deluge must come. And the last statement becomes a keystone in the arch of testimony; for every man must know that such a fund of waters could not have existed without pointing to a deluge to come! Did a deluge come? Independently of every other consideration, I am bold to say, if the earth has not been deluged again and again, then every leaf of the geologic record is a lie; then the molten earth has no conclusion, there can be no fires in the universe, no suns or flame; for law is law in every nook of creation, and if solid matter fell to the earth, and formed its mass, so did its waters fall upon it as the last remnants of annular matter. With this mass of evidence pointing to a deluge, we will next see how true to these indications there came a terrible fall of waters from the "great deep" on high.

CHAPTER VI.

THE NOACHIAN DELUGE.

A professor from one of the first institutions in this country once declared, as he no doubt conceived to the discredit of biblical history, that "no one but a D.D. now believed there ever was a deluge." It was well said! To the deathless honor of the "D.D.'s" may it always be said, they stand for the testimony and the law! I have shown in the foregoing chapter that every feature and phenomenon of the Adamite age point to a future deluge as an utter and absolute necessity. Let the reader re-survey the statements made in reference to the heavens and the earth, the divisions of waters, the "stars," and the "lighters"; the light of the first, second, third days; the day of rest; the Eden world and its climate; the rainless period, when the whole surface of the earth was watered by a mist only;* man's longevity; the "giants of those days;" the absence of the rainbow, etc., etc.,

* "But there went up a mist from the earth and watered the whole face of the ground." (Gen. 11: 6.) I understand that by this process the whole earth was watered. It must seem to him who critically examines this statement, that there is another philosophic sequence of the ruling of upper waters in that long day of physical rest. As the earth rotated, one-half of it must have been more directly under the influence of diffused solar heat for half the time; during which, one-half of the atmosphere would absorb aqueous matter from sea and land, and during the remainder of the time the atmosphere being carried by rotation somewhat beyond solar influence into the "cool of the day," would water the earth with excessive dews or mist. Yet there are strong reasons for claiming that one-half the day presented a scene of rising fog, and the other of descending mist. But we must always bear in mind that annular vapors were continually saturating the air on high, and would thus add to descending "mists."

and if possible draw any other conclusion than that this earth, in the ordering of nature, was fated for a coming fall of water. What is more natural? And what natural visitation could be more appalling, far-sweeping and destructive, than this inevitable dispensation?

I presume the biblical narrative of the flood is in the possession of every reader of these pages and I will therefore not insert it here.

The first impression given to the reader of the Mosaic account is the universality of the falling waters, which of course necessitates an annular source.

Can the philosophic mind, as it contemplates this great world of law, conceive of any source in the ordering of the God of Nature, from whence such a stupendons downfall of waters could come, other than this most natural one? A rain from the mighty "deep" alone could thus have swept the earth. And when we contemplate that there is a volume of water now on the earth, and in its rocky frame, sufficient to make a thousand terrific deluges,—every one of which could drown the world of living beings, and which has fallen to the earth, necessarily as stupendous cataclysms,—how can we reasonably expect that this historical and traditional narrative can refer to any other than the closing scene of annular declension? Let us reduce the extent of this great debacle of waters to the lowest minimum this narrative will allow, then take into consideration the well-known fact that there scarcely is a nation, tongue or tribe on earth that has not a tradition of this great event, and yet we will fail to find any existing source of such a rain. Study the biblical account of the flood, and tell me, did that rain descend from the clouds?

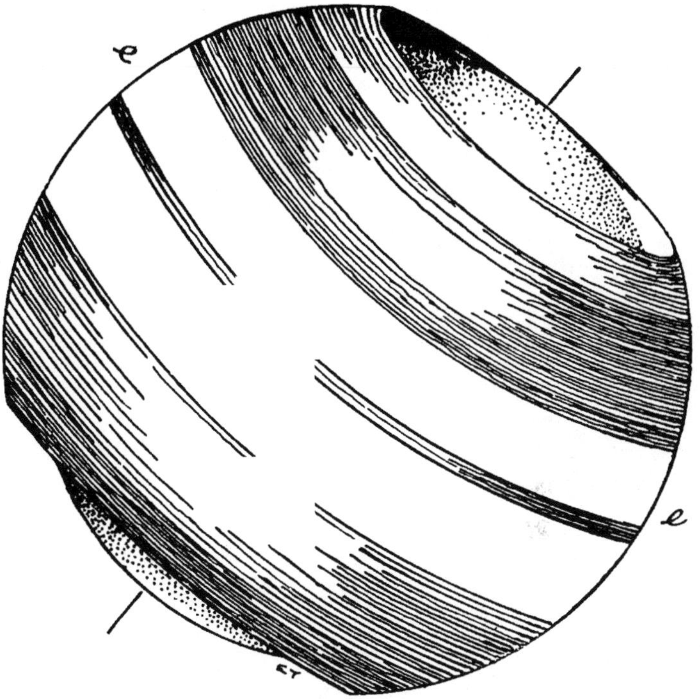

Fig. 5. THE LAST CANOPY OF EARTH.

Here is our *Last Earth Canopy*. It has banished the last ice period, and the Eden earth blooms again. Biblical and legendary man dwells naked in a warm and genial world. The human family, the world over, for unknown time look up to a *watery heaven* and give it a name signifying that condition. The Hebrews called this heaven *Shamayim*, "there waters"; the Greeks called it *Ouranos*, "water heaven"; the Hindus called it *Varuna*, "water heaven"; the Latins called it *Cœlum*, and this, too, was a watery heaven, for it *passed away*. So did the heaven of the ancient Egyptians, Japanese, Scandinavians and Mexicans.

But the Last Earth Canopy must fall. It opens at the equator (*e*, *e*), and the vapors slowly float to the poles, and begin to fall. Again, as in ages gone by, snows begin to chill the earth. The sun shines in upon the equatorial earth through the opening and air in that region rises. This starts air currents from the poles and these currents bear the falling vapors back toward the equator, thus making one long-continued downpour of waters in medial latitudes. So, in the windup of canopy influences, there must be not only a vast accumulation of snows at the poles, but long-continued and devastating floods in warmer lands.

Those snows now cap the poles. Those floods have sent their immortal witnesses down the ages, and they speak from the sacred pages of our fathers and in the songs and legends of the whole earth. The last canopy having fallen, heaven has made a rew and eternal covenant with earth, and the Bow is the eternal Testator.

Is it not a demonstrable fact, that if the clouds were its source, eternal law was suspended? Is this the order of nature? Is this the administration under which worlds are evolved? We cannot admit the fraction of law in the universe of God! But if the deluge did not come from the clouds, then it came from beyond, or above the clouds. And again we are compelled to call in the annular waters as the only competent source. Let us analyze the account a little more minutely. Is there any intimation in the narrative itself that the Noachian rain did not come from the clouds? There certainly is, not only an intimation, but unmistakable and positive declarations, that the clouds did not and could not have supplied such a rain. Gen. 7: 11 tells us that rain came from a source that was "broken up" at the time the waters fell; that it came from the *cataractae* of heaven—from the "fountains of the great deep," through the imaginary "windows of heaven."

If we will but reflect that at that time mankind believed that there was a great deep on high, from which all rains descended, that the Deity resided in that part of heaven and presiding over its fountains, watered the earth through windows opened for that purpose, we cannot avoid the conclusion that the sacred pen described this great event, true to the indications, and in absolute harmony with facts. But the annular theory demands the same conclusion. It requires that the source of the deluge should have been "broken up" at that very time, for it does not now exist. And any one can see, that if that rain came from beyond the clouds, it came from revolving waters or vapors; and also that no fountains or source of floods could have been "broken up," except such a source. A little thoughthere must settle this question in the philosophic mind.

Again, we are told over and over, in the eighth and ninth chapters of Genesis, that there will be such floods no more forever. Then it is impossible for such rains to occur again, and then we are forced to admit again, that the source *has* been " broken up." But no source of floods can be broken up but the source of annular floods! If the " fountains of the deep " were on the earth or in the " seas," then they are not " broken up." If they were in the clouds, they are not. If that source has not been destroyed we are under the same precarious reign of floods still; and no physical assurance whatever protects us from their recurrence. Thus, according to the annular hypothesis, the declaration is positive and unmistakable, that man is forever safe from a deluge; that the waters can "no more become a flood " to " destroy all flesh "—the very same declaration made by the historian, and which certainly has no significance except in this light.

Again, the annular theory declares to all races of men under heaven that an eternal covenant is made between them and their Creator, and that the rainbow is an everlasting token of the same, just as the biblical account maintains. The two must agree, as they must in every particular, for both are the voice of nature. These wondrously harmonious facts! What marvelous truths unfold to view in the resolution of these once mysterious statements! What stronger evidence can vindicate a theory! They are the adamantine sills upon which the true theory of creation is planted forever.

Now we know that a devastating flood did visit the

earth in the human period, and we know its all-competent and philosophic source. "In the 600th year of Noah's life, in the second month, on the 17th day of the month, the same day were all the fountains of the great deep broken up, and the windows of heaven were opened. And the rain was on the earth forty days and forty nights."

It seems to me there cannot be a man of reason who cannot see in this declaration annular-canopy waters emphatically portrayed. For what reason have men concluded that the "great deep" here mentioned was the great ocean-fund on the earth? Why should the imaginary "windows of heaven" be opened to let down the waters, if those waters were located in the fountains in the earth? Let us imagine ourselves placed in the same situation as the ancient human race, fully believing there were fountains of waters on high, over which the Deity presided, and which were the sources of descending floods or rains. Then let us see the same heavens they saw, cleared of upper vapors,—veritable fountains of the deep. Would we not under the same circumstances see that those sources of falling waters "were broken up?" And further, knowing that the bow could not be seen if any of those vapors remained on high, and seeing it painted on the skies, as an actual sign that such vapors had all fallen to the earth, would we not have said, "No more deluges can occur;" and this, the bow, is a token of the same? Would we not have penned an account of this dispensation, just as the ancient historian did? Unless we admit this as the true rendering of this wonderful narrative, what can we make of it? What reason, what philosophy can be found in it?

Do we not know full well that no such terrific rains could possibly come, in the order of nature, from any other source than from that beyond the clouds, where inexorable law put the fountains of all descending waters in primitive geologic times; and whence all the waters now on the earth must have fallen? The idea that the great deep of Genesis was or is the terrestrial ocean, is a post-diluvian one, and necessarily arose from the fact that our oceans, after the upper one had fallen, became the only one man saw. Could I place before my readers the vast array of facts that may be drawn from the antediluvian and interdiluvian mythologies in support of the universal belief that the heaven, the home of deities, was a region of abundant rivers and fountains, of oceans traversed by golden ships, etc., they would never doubt the Hebraic idea that all rains were given by Jehovah of the Gods, drawn from celestial fountains, and poured down upon the earth through the "windows of heaven." The oldest Hebraic histories teem with this idea. The great Psalmist says, "Praise Him ye waters above the heavens," "Deep calleth unto deep at the noise of thy water-spouts." Mankind formerly believed that the clouds of post-deluge times were filled from a great fund of waters above them; and this was especially so among the Hebrews. (See Job 26: 7-14, and 28: 11, 24, 26; also 36:30, and 38: 8-26.)

This idea runs through the works of all the ancient poets. Homer's and Virgil's writings reveal it on almost every page. The great deep of the Hebrews was the same as the Okeanos (Oceanus) of the Greeks and the celestial Nile of the Egyptians. When the great abyss fell, its name was translated from the skies to the waters of the earth. In a succeeding volume of this series these facts will be fully established.

Unless we adopt the philosophic view here set forth, what meaning can the forty days' rain be made to assume? Waters were sent from a great deep through the windows of heaven. Reason forever refuses to entertain the idea that a rain from the clouds met waters from the terrestrial oceans, and both combined overran the earth. It is unphilosophic and unnatural. See to what endless disorder and confusion, contradiction and self stultification, such an idea impels us. It at once tells us that the bow as a token or sign of security means nothing, and running from sequence to cause it could be readily proven that the first and eighth chapters of Genesis are a tissue of contradictions and falsehoods. But with the fountains of the great deep placed on high,—the veritable "waters above the firmament,"—we can readily understand why the "windows of heaven were opened," and why "all the fountains were broken up." A grander and more significant truth was never penned by the hand of man. All those fountains were broken up at that time,—the very fact which the rain-bow proclaims forever, around the circuit of the earth. Oh, when will the master-minds of the world grasp this momentous idea! What a sad spectacle the unbelieving world presents to-day, simply because the doctrine of a universal deluge is denied! Now, I do not expect to claim that the great deluge in its might and fearful magnitude and grandeur, swept the entire earth, but that its appalling effects were felt in some form in every part of the globe cannot be denied. In another chapter I will give solid reasons for claiming that the oceans stand to-day vastly deeper than they did in the Adamite age, and I will let the most skeptical man answer the following question. If the fountains of the great deep should, during a down-rush of water for six months, raise the ocean so as to cause its level to climb upon the continents, is it likely that many individuals of any race of terrestrial beings, could survive the catastrophe? And would not such a rain or flood convince the survivors of the same that it was universal? All I claim, and ask, is that men admit that a universal flood is no unnatural thing. No man of reason will for one moment doubt that the earth was deluged universally when the oceans fell to its surface. The oceans did fall to its surface! And what philosophy can there be advanced against the claim that some part of the oceans fell in the days of Noah, since we all know they did fall some time? If men choose to say they all fell in pre-archæan times, I choose to say they could not and did not, and that all the evidence is on my side.

What consternation would fill the mind of humanity, if from some exhaustless fund of waters, the earth should in this age be again compelled to undergo such a baptism! Such a rush of waters as would drench the hills for years, perhaps ages, fill the valleys till they leap their boundaries, and pour into the ocean from millions of river-mouths, till its level would rise even one foot, would literally drown the earth; and a few thousand years would obliterate much of the impression from the human mind, and all physical appearauces of its track. If at such a time, that fund of waters were exhausted by the down-rush, and man in ages to come could see no longer any philosophic cause of deluges, and forgetting that a source of such floods did once exist, he would begin to doubt the truth of the old histories and traditions relating thereto.

This is the exact condition to-day of our knowledge of the last great deluge. The history of the event is chronicled in the oldest records of the races. Its source has vanished, and men doubt that it ever occurred. But let me again repeat: the evolution of the earth demands that such a source be supplied. As man directs his mind to this investigation he must and will supply it. So that if every trace of the history, tradition or physical appearance of such an event be utterly lost, man must and will conclude that the earth has been deluged many times. The oceans, as they roll around the planet, are the aggregate of almost endless additions of water during the ages.

At this point let us look back upon the ground we have left. See the order in which this remarkable account is given. We are told there was a day when it did not rain on the earth. Surely every one of my readers can see that this necessarily excludes the sun's direct heat from the earth's surface. The sun must shine upon the earth, and heat its surface before air-currents can arise and enter upon their round. It is a commingling of air-currents of different temperatures that causes rains to fall. Consequently no currents, no rain; and no rains, no sun. But no sun necessitates upper vapors, and upper vapors necessitate an annular appendage; an annular appendage of vapors necessitates a deluge; and a deluge from that source means the clearing of the skies, and the advent of the bow; and the clearing of the skies, necessitates a great "wind," and the beginning of the grand air-currents of the atmosphere; and this, so long as it continues, necessitates the regular alternation of seasons. If there be anything at all in physical law, we certainly cannot avoid this conclusion. How did it ever happen, that the author or authors of Genesis related these facts in such harmonious accord with all those conditions which an annular arrangement of waters necessitates? How did it ever happen that the "mistakes of Moses" were all made in the line of eternal law? "Mistakes !" Facts related for what? To establish a theory which the least variation or contradiction would vitiate? We must value that history, and that historian, that presents what have puzzled the greatest minds of earth four thousand years.

Again, we are told there was light in the terrestrial heavens before the sun appeared; but light before the advent of the visible sun necessitates a fund of intercepting matter above the terrestrial firmament; and this requires an annular or belted canopy; and this requires a day of physical rest, a rainless age, an Eden clime, and long life; and (may I not also claim?) a race of giants.* (See Genesis 6: 4.) But each and all these things demand a suspension to a great extent of the regular alternation of seasons, seed-time and harvest, cold and heat, and day and night, before the deluge. And this condition of things before the deluge demands a change and a regular alternation of the same, as the waters fell and the heavens became clear. Hence the

* The presence of upper vapors, entering the atmosphere on their way to the earth, via the polar regions, necessitates an atmosphere of greater buoyant power. (For much of the weight of all the waters of the deluge was added thereto, so long as it existed in the atmosphere, which may have been the case for millenniums.) And a greater buoyant power necessitates larger bodily frame. Hence the greater size of antediluvian animals. Question: Could there have been ' giants in those days " if there had been no upper waters?

declaration made immediately after the waters fell. (See Gen. 8: 22.) I need not tell the reader how nearly these natural causes and sequences flow in harmony with the demands of the earth's annular system.

In order to prove more fully the claim that the great deep of diluvian times was the upper ocean of ancient man, we will draw a little from traditions, that yet, after thousands of years, live in the human mind. As we proceed, let the reader notice the evident fact that man, in the infancy of the race, and the rudeness of his intellect, coupled the deluge with celestial streams, celestial deities and celestial monsters guarding celestial fountains. It is not necessary for me to call the attention of classical students and thinkers to these facts. They know the pages of mythological literature are replete with these ideas; and every man must admit the only claim I build upon this evidence, that when these thoughts were penned, these ideas pervaded the human race; that one main thought runs through all ancient traditional lore; that the "great abyss" of the Hebrews; the Oceanus of the Greeks and Romans; and the Nilus of the Egyptians were the "waters above the firmament,"—the earth's canopy appendage.

CHAPTER VII.

LEGENDS OF THE FLOOD.

Having conclusively shown that vast and terrific deluges have been a philosophic necessity from the remotest geologic ages, and having shown that the peculiar testimony of Genesis can mean nothing less than a recital of the effects of the fall of the last remnant of the earth's annular system, I will here append a brief chapter of the "*F*lood Legends," as they have existed in the history and memory of the human race for unknown ages. I do this, not with the mere aim to strengthen the foundation already laid, but also to present some facts of interest, permitting one to draw his own conclusion as to the bearing they have upon the theory I have advanced. Yet it is impossible not to see in many of these legends the intimate relation between cataclysms and their efficient and natural cause. First let us take a philosophic glance at the value of these legends.

Such widespread desolation must have indelibly impressed the human mind, and inasmuch as the account has come down to us through the custodians of the most reliable history,—the guardians of civilization, the Aryans, Phœnicians, Greeks and Hebrews, it is no difficult task to co-link even the rudest form of flood traditions with the one terrible visitation so graphically related by the ancient penman. Its shadows will never pass from the historic page. Men may impugn and ridicule the narrative. Yet the fact remains, that a self-sustaining history is there; and the combined sophistry of all time cannot shake it. The day will come when even the most incredulous will admit the main truth recorded from the very fact that it is self-corroborative. Let the reader again peruse the plain unvarnished narrative as recorded in Genesis. We are indebted to Berosus, who is supposed to have been a Chaldean priest, for the most valuable traditional account of the flood. He lived some time in the third century B. C., and seems to have had access to the Babylonian records. Some of these, including the flood legend, he translated into the Greek language. This latter I give as translated from the Greek historian, and is as follows:—

"After the death of Ardates, his son Xisuthrus (Noah) reigned 18 sari. In his time happened a great deluge, the history of which is thus described. The god Chronos appeared to him in a vision, and warned him that upon the fifteenth Dæsius there would be a flood, by which mankind would be destroyed. He therefore enjoined him to write a history of the beginning, procedure and conclusion of all things, and to bury it into the City of the Sun at Sippara; and to build a vessel, and take with him into it, his friends and relations, and to convey on board every thing necessary to sustain life, together with all the different animals, both birds and quadrupeds, and trust himself fearlessly to the deep. Having asked the deity whither he was to sail, he was answered 'To the gods'; upon which he offered up a prayer for the good of mankind. He then obeyed the divine admonition and built a vessel, five stadia in length, and two in breadth.* Into this he

* One stadium = 625 Roman, 600 Greek, and 609¾ English feet.

put every thing which he had prepared, and last of all conveyed into it his wife, his children, and his friends. After the flood had been upon the earth, and was in time abated, Xisuthrus sent out birds from the vessel, which, not finding any food, nor any place whereupon they might rest their feet, returned to him again. After an interval of some days, he sent them forth a second time, and they now returned with their feet tinged with mud. He made a trial a third time with these birds, but they returned to him no more. From which he judged that the surface of the earth had appeared above the waters. He therefore made an opening in the vessel, and upon looking out, found that it was stranded upon the side of some mountain, upon which he immediately quitted it with his wife, his daughter and the pilot. Xisuthrus then paid his adoration to the earth, and having constructed an altar, offered sacrifices to the gods, and with those who came out of the vessel with him, disappeared. They who remained within, finding that their companions did not return, quitted the vessel with many lamentations, and called continually upon the name of Xisuthrus. Him they saw no more; but they could distinguish his voice in the air, and could hear him admonish them to pay due regard to religion; and likewise informed them, that it was on account of his piety, that he was translated to live with the gods, and that his wife and daughter had obtained the same honor. To this he added, that they should return to Babylonia, and as it was ordained, search for the writings at Sippara, which they were to make known to all mankind. Moreover, that the place wherein they then were, was the land of Armenia. The rest, having heard these words, offered sacrifices to the gods, and taking a circuit, journeyed towards Babylonia. The vessel being thus stranded in Armenia, some part of it yet remains in the Gordyan mountains."

Such is an account of the last grand debacle of exterior waters, as it comes to us from the historian of the Chaldees. It bears upon its face some important and undeniable truths. 1st. That its original source and that of the Biblical account were one and the same. 2d. That the former had been long preserved in the mind and memory of a different nationality, or people. The tradition of the Chaldees shows that its custodians were a maritime people, one accustomed to the waters. Their ark was a ship, and built for the ocean, and their Noah was commanded to launch fearlessly upon the "deep." The Biblical ark was a thebotem, a chest or box, and every term used in the history of the same, points to the fact that it was a place of refuge for an inland people. The Chaldees had a pilot, a term and personage employed only among inhabitants of the waters. 3d. It shows that there was a vast lapse of time, from the date of the deluge to the time when the account was placed upon the Chaldean records. So long was it that this people claimed it as their own history, just as every race and tongue, having a similar tradition, does to-day. It is a perfectly natural result. Each tribe and race has perpetuated the knowledge of the deluge in its own language; each has its own Noah, ark, ship or canoe. It would be exceedingly strange and unphilosophic if it did not. There is another version of the Chaldean account a little different: "The deity, Chronos, foretold to him (Sisithrus), that on the fifteenth day of the month Dæsius, there would be a deluge of rain, and he commanded him to deposit all the writings whatever, which were in his possession, in the city of the Sun at Sippara. Sisithrus, when he had complied with these commands, sailed immediately to Armenia, and was presently inspired of God. Upon the third day after the cessation of the rain, Sisithrus sent out birds by way of experiment, that he might judge whether the flood had subsided. But the birds passing over an unbounded sea, without finding any place of rest, returned again to Sisithrus. This he repeated with other birds, and, when upon the third trial he succeeded (for the birds then returned with their feet stained with mud), the gods translated him from among men. With respect to the vessel, which yet remains in Armenia, it is a custom of the inhabitants to form bracelets and amulets from its wood."

I wish here to again call the attention of the reader to the indisputable fact, that the continual decline of revolving vapors, as they progressed toward the poles, would cause them to grow thin near the equator, and that the sun became visible in the equatorial world long before the great catastrophe to which these traditions allude. This seems to have been the case of the cherub sun more than a thousand years before the days of Noah. That the sun's coming into view would be the only physical means, possessed by the human race, as a warning against the impending danger. How significant then the fact that Chronos, the god of time, the vapor-veiled sun, is represented here as announcing to the human race the coming deluge? After this great cataclysm of water, when the sun became clear and visible to the entire world, what could be more natural and reasonable than that the remnant of the human race, as they multiplied and filled the earth, knowing that this luminary came into view as the waters fell, should look upon it as a deity—a measurer of time—and finally, as a veritable personage who had given warning of the coming deluge. Now, anyone can see to-day that the sun would be to the scientists of this age, an actual measurer, and he would not be very much of a mathematician, at this age, who could not by noticing the appearance of the moving belts, the perpetual change in the halo, the fitful and frequent falls of vapor, as the years rolled by, calculate the month and the day of the final collapse. It seems that man knew of this coming dispensation long enough to enable him to build an ark. How did he get the information? The simple fact is, that Noah was informed of God, either as a seer or mathematician. Which?

As it may be of some interest to the reader to learn something of the records from which this tradition came, I will present in this chapter some parts of the rude legend of the flood, imprinted on brick tablets, perhaps in the early age of the Babylonian monarchy, nearly, if not quite, 4,000 years ago, and stored away in the libraries of Nineveh, and other cities, now mouldered to dust and marked only by rounded heaps or mounds. From these mounds, the persistent efforts of Layard and Smith have brought to light vast numbers of these tablets, veritable books of those long-lost ages, which have been so far deciphered and translated, as to show conclusively what they are. These volumes are inscribed in cuneiform characters,—characters so exceedingly old that it was but a happy accident that the key to their meaning was discovered and that archæologists are now able to interpret them. These tablets are so mutilated and broken that their full text cannot be made out.

"I listened to the decree of fate that he announced, and he said to me:—'Man of Shurippak, son of Ubar-atutu—thou, build a vessel and finish it (quickly).——(By a deluge) I will destroy substance and life.—Cause thou to go up into the vessel the substance of all that has life.——The vessel thou shalt build—600 cubits shall be the measure of its length—and 60 cubits the amount of its breadth and of its height.——(Launch it) thus on the ocean, and cover it with a roof.'——I understood, and I said to Ea, my lord:—'(The vessel) that thou commandest me to build, thus—(when) I shall do it,—young and old (shall laugh at me.)'——(Ea opened his mouth and) spoke.—He said to me, his servant:—'(If they laugh at thee) thou shalt say to them:—(shall be punished) he who has insulted me, (for the protection of the gods) is over me— . . . like to caverns . . . — . . . I will exercise my judgment on that which is on high, and that which is below . . . — . . . Close the vessel. . . — . . . At a given moment that I shall cause thee to know,—enter into it, and draw the door of the ship toward thee.—Within it, thy grains, thy furniture, thy provisions,—thy riches, thy men-servants and thy maid-servants, and thy young people—the cattle of the field, and the wild beasts of the plain that I will assemble—and that I will send thee, shall be kept behind thy door.' Khasisatra opened his mouth and spoke;—he said to Ea, his lord: 'No one has made (such a) ship.——On the prow I will fix . . . —I shall see . . . and the vessel . . . —the vessel thou commandest me to build (thus)—which in . . .' On the fifth day (the two sides of the bark) were raised.—In its covering fourteen in all were its rafters—fourteen in all did it count above.—I placed its roof, and I covered it.—I embarked in it on the sixth day; I divided its floors on the seventh; I divided the interior compartments on the eighth. I stopped up the chinks through which the water entered in;—I visited the chinks, and added what was wanting.—I poured on the exterior three times 3600 measures asphalte,—three times 3600 measures of asphalte within.—Three times 3600 men, porters brought on their heads the chests of provisions.—I kept 3600 chests for the nourishment of my family,—and the mariners divided among themselves twice 3600 chests.—For (provisioning) I had oxen slain;—I instituted (rations) for each day.—In (anticipation of the need of) drinks, of barrels, and of wine—(I collected in quantity) like to the waters of a river, (of provisions) in quantity like the dust of the earth.—(To arrange them in) the chests I set my hand to.— . . . of the sun . . . the vessel was completed.—— . . . strong and—I had carried above and below the furniture of the ship—(This landing filled the two thirds.) All that I possessed I gathered together; all I possessed of silver I gathered together—all that I possessed of gold I gathered together—all that I possessed of the substance of life of every kind I gathered together,—I made all ascend into the vessel; my servants, male and female,—the cattle of the fields, the wild beasts of the plains, and the sons of the people, I made them all ascend. Shamash (the sun) made the moment determined, and—he announced it in these terms:—'In the evening I will cause it to rain abundantly from heaven; enter into the vessel and close the door'—the fixed moment had arrived, which he announced in these terms;—'In the evening I will cause it to rain abundantly from heaven.'—When the evening of that day arrived, I was afraid—I entered into the vessel and shut my door.—In shutting the vessel, to Buzur-shadi-rabi, the pilot,—I confided this dwelling with all that it contained. Mu-sheri-ina-namari—rose from the foundations of heaven in a black cloud;—Rammon thundered in the midst of the cloud, —and Nabon, and Sharru marched before;—they marched, devastating the mountain and the plain;—Nergal, the powerful, dragged chastisement after him; —Adar advanced, overthrowing before him;—the archangels of the abyss brought destruction.—in their terrors they agitated the earth.—The inundation of Rammon swelled up to the sky,—and (the earth) became without lustre, was changed into a desert. They broke . . . of the surface of the earth like . . . ;—(they destroyed) the living beings of the surface of the earth. —The terrible (deluge) on men swelled up to (heaven). —The brother no longer saw his brother; men no longer knew each other. In heaven—the gods became afraid of the water-spout, and—sought a refuge; they mounted up to the heaven of Anu.—The gods were stretched out motionless, pressing one against another like dogs.—Ishtar wailed like a child,—the great goddess pronounced her discourse:—'Here is humanity turned into mud, and—this is the misfortune that I have announced in the presence of the gods.—So I announced the misfortune in the presence of the gods,—for the evil I announced, the terrible (chastisement) of men who are mine.—I am the mother who gave birth to men, and—like to the race of fishes, there they are filling the sea;—and the gods, by reason of that—which the archangels of the abyss are doing, weep with me.' —The gods on their seats were seated in tears,—and they held their lips closed, (revolving) future things. Six days and as many nights passed; the wind, the water-spout, and the diluvian rain were in all their strength. At the approach of the seventh day the diluvian rain grew weaker, the terrible water-spout—which had assailed after the fashion of an earthquake, grew calm, the sea inclined to dry up, and the wind and the water-spout came to an end. I looked at the sea, attentively observing, and the whole of humanity had returned to mud; like unto sea-weeds the corpses floated. I opened the windows, and the light smote on my face. I was seized with sadness; I sat down and I wept;—and my tears came over my face. I looked at the regions bounding the sea; toward the twelve points of the horizon; not any continent.—The vessel was borne above the land of Nizir,—the mountain of Nizir arrested the vessel, and did not permit it to pass over. —A day, and a second day the mountain of Nizir arrested the vessel, and did not permit it to pass over;—the third and fourth day the mountain of Nizir arrested the vessel, and did not permit it to pass over;—the fifth and sixth day the mountain of Nizir arrested the vessel, and did not permit it to pass over. At the approach of the seventh day, I sent out and loosed a dove. The dove went out, turned, and—found no place to light on, and it came back. I sent out and loosed a swallow; the swallow went, turned, and—found no place to light on and it came back. I sent out and loosed a raven; the raven went and saw the corpses on the water; it ate, rested, turned, and came not back. I then sent out (what was in the vessel) toward the four winds, and I offered a sacrifice. I raised the pile of my burnt-offering on the peak of the mountain; seven by seven I disposed the measured vases,—and beneath I spread rushes, cedar, and juniper-wood. The gods were seized with the desire of it —the gods were seized with a benevolent desire of it; —and the gods assembled like flies above the master of the sacrifice. From afar, in approaching, the great goddess raised the great zones that Anu has made for their glory (the gods). These gods, luminous crystal before me, I will never leave them; in that day I prayed that I might never leave them.

"Let the gods come to my sacrificial pile!—but never may Bel (sun) come to my sacrificial pile! for he did not master himself, and he has made the water-spout for the Deluge, and he has numbered my men for the pit. From afar, in drawing near, Bel—saw the vessel, and Bel stopped;—he was filled with anger against the gods and the celestial archangels:—'No one shall come out alive! No man shall be preserved from the abyss!'—Adar opened his mouth and said; he said to the warrior Bel:—'What other than Ea should have formed this resolution?—for Ea possesses knowledge, and (he foresees) all.'—Ea opened his mouth and spoke; he said to the warrior Bel:—'O thou, herald of the gods, warrior,—as thou didst not master thyself, thou hast made the water-spout of the Deluge.—Let the sinner carry the weight of his sins, the blasphemer the weight of his blasphemy.—Please thyself with this good pleasure, and it shall never be infringed; faith in it never (shall be violated). Instead of thy making a new deluge, let lions appear and reduce the number of men;—instead of thy making a new deluge, let hyenas appear and reduce the number of men;—instead of thy making a new deluge let there be famine, and let the earth be (devastated);—instead of thy making a new deluge, let Dibbara appear, and let men be (mown down). I have not revealed the decision of the great gods;—it is Khasisatra who interpreted a dream and comprehended what the gods had decided.' Then, when his resolve was arrested, Bel entered into the vessel.—He took my hand and made me rise.—He made my wife rise, and made her place herself at my side.—He turned around us and stopped short; he approached our group.—' Until now Khasisatra has been made part of perishable humanity;—but lo, now Khasisatra and his wife are going to be carried away to live with the gods,—and Khasisatra will reside afar at the mouth of the rivers.'—They carried me away and established me in a remote place at the mouth of the streams," etc., etc.

Here we have in rude form the legend of the flood lettered on brick and stone thousands of years ago. In the libraries of Nineveh, a city within whose walls were "six score thousand persons who knew not the right hand from the left," they were stored away and when her temples went to dust there they remained. Amid the shock of war her walls went down, and the shout of armies echoed through her streets. When citadels and temples and courts had gone to dust, her silent libraries were waiting but to speak in other tongues, in a world of light. That light came and her dust awoke and she speaks again. She says in language too plain to be mistaken, that when these rocky volumes were inscribed, the tradition then was old! So old the carver or printer scarcely knew the doubtful from the true.

Some of my readers perhaps are aware that some that aim to cast discredit upon the Biblical account of the deluge, make the claim that the writer of Genesis borrowed the account from the Nineveh columns. How strange that any man of reason should make the claim. Place the two side by side, and tell me which is more natural, which is more in accord with facts, with law? Which more distinctly portrays a down-rush of exterior floods; which possesses the more harmonious links of truth in the light of modern discoveries; which of the two reveals more plainly the necessity of waters beyond the firmament? A Sabbath of physical rest, a rainless age, an Eden world, a windless, nightless, winterless age? Which of them reveals the positive fact that there can never be another deluge; and which of them seals the eternal covenant with the stamp of God impressed upon the clouds? How many of these were borrowed from Nineveh's sapient piles? While these ancient tablets are eugaging our attention, let me call the reader's thoughts to the patent fact that their authors, like the rest of the older world, looked upon the heavens as the home of fountains and rivers. The imaginary lands of the gods beyond the clouds were a world of waters traversed by celestial ships inhabited by all imaginary monsters.

Here the sage who escaped the flood and was translated to the skies relates to Izdubar the story of the flood. Izdubar seeks an entrance into the celestial world and pleads in behalf of his dead companion who is resting uncomfortably in Hades, and asks that he be rescued. His dead companion also joins in the appeal in the following language: . . . " Return me from Hades, the land of my knowledge, from the house of the departed, the seat of the god Irkalla from the house within which there is no exit. From the road the course of which never returns. *F*rom the place within which they long for light. *T*he place where dust is their nourishment and their food mud. Its chiefs also like birds are clothed with wings. Light is never seen; in darkness they dwell." Such was their picture of Hades, the under-world, and their description of the place proves their belief in such a place and a future life. But now comes the picture of the land of the blessed after death. . . . " Return me to the place of seers which I will enter . . . treasured up a crown; . . . wearing crowns who from days of old ruled the earth. To whom the gods, Anu and Bel, have given renowned names. A place where water is abundant drawn from perennial springs." Also let us note in this connection one more very significant fact, that the ancient Babylonians considered the sun to be the author of the deluge. On one column are inscribed the sentiments of Noah, " May the gods come to my altar; may Bel (the sun-god) come not to my altar; for he did not master himself and made a deluge, and my people he had consigned to the deep. From of old also Bel in his course saw the ship, etc." Again when the patriarch of the ship was about to leave his vessel, he says, "Adrahasis (Noah,) a dream they sent and the judgment of the gods he heard. When his judgment was aecomplished Bel went up to the midst of the ship; he took my hand and raised me up; he caused to raise and bring my wife to my side . . . When Hasisadra and his wife, and the people to be like gods were carried away, then dwelt Hasisadra in a remote place at the mouth of the rivers." Thus it seems the idea of celestial rivers and streams "fed by perennial fountains" was a common one 4,000 years ago. And that Bel (Belus), the Sun, was Nineveh's god, and her author of the flood. How natural these things appear! The sun coming into view as the waters fell, was esteemed a deity, and the direct cause of man's destruction. Hence for unknown generations, he was feared and adored by the survivors of the flood. This will be more fully understood in connection with other traditions. I shall treat elsewhere of the ancient belief of mankind that the heavens were supported by giants, whose heads received the lofty vault of the firmament, and whose feet were planted in the depths of the earth; that the origin of this belief was the actual existence in the eastern and western skies, of the appearance of vast columns rising from the horizon, and spreading out against the face of the sky, their huge Briarean arms, the actual vapor-bands that afterwards fell. Now hear what George Smith, the indomitable searcher and discoverer of these old tablets, says after years of study in this direction. "They (ancient Babylonians) held the idea that at a little distance from them there were giants who controlled the rising and setting sun, and that the orb of day was looked after, and sent on its course by these beings, who had their feet in the lower region of hell while their heads touched and probably upheld the heavens." Veritable pillars of Hercules,—pillars of the sun. How immortal are some crude ideas. This idea which must have obtained before the deluge, lived in the mind of men for unknown ages after these phenomena disappeared.

When America was discovered there existed among the Mexicans a tradition of a deluge, which represented that a couple of people were saved therefrom in a ship or raft, from which birds were sent to ascertain whether the waters had subsided. Some of these it is stated saw the floating carcasses on the water and fed thereon. Humboldt tells us that "of the different nations who inhabit Mexico, paintings representing the deluge are found among the Aztecs, the Mizletecs, the Zapotecs, Tlascaltecs and the Mechoachans. The Noah, Xisuthrus or Menu of these nations is Coxcix Teocipactli or Tezpi. He saved himself and his wife in a bark, or, according to other traditions, on a raft. But according to the Mechoachans he embarked in a spacious 'acalli,' with his wife, his children, several animals and some grain, the preservation of which was important to mankind, when the great spirit ordered the waters to withdraw. Tezpi sent out from his ship a vulture, the Zapilote. This bird that feeds on dead flesh did not return. . . . Tezpi sent out other birds, one of which, the humming bird, alone returned, holding in its beak a branch covered with leaves. Tezpi, seeing that fresh verdure began to clothe the soil, quitted his bark near the mountain of Colhauacan" (Humb. Res. p. 65.)

Another tradition of the ancient Mexicans states that 4,800 years after the creation, a great inundation took place; that before this the country of Anahuac was inhabited by giants; that after the deluge the survivors built a hill in the shape of a pyramid, the top of which was to reach the clouds. This displeased the gods, who hurled fire on the builders, some of whom were killed, and the monument was afterwards dedicated to Quelzolcoatl (Jupiter). Thus we see in the legends of Mexico, parts of the Chaldean tradition, such as the floating carcasses and the devouring birds—traditions which were buried for 4,000 years in Babylonian libraries. During all this time, then, the memory of this great dispensation lived in the mind of man. How indelible the stamp it has placed upon the deathless pages of tradition.

Ellis, in his "Polynesian Researches," says: "The Sandwichers believe that the Creator destroyed the earth by an inundation that covered the whole earth except Mouna Roa in Hawaii, on the top of which one single pair had the good fortune to save themselves." Thus in the midst of the Pacific Ocean, cut off from every source of information from the scenes of Armenian history, the same tradition lives. Another thing plainly to be seen in these deluges is the cataclysmic character of the devouring waters. "The archangel of the (celestial) abyss brought destruction." "The waters rose to the sky." The inhabitants of the earth "fill the sea like fishes." Their "corpses floated like sea-weeds." It was a war of elements most terrific. "Water-spouts pouring from the abyss." Ea opened his mouth and spake; he said to the warrior Bel (the sun), "Oh, thou, herald of the gods, warrior—as thou didst not master thyself (didst not consider), thou hast made the water-spout of the Deluge." Again—"Six days and as many nights passed; the wind, the water-spout, and the diluvian rain, were in all their strength." "In heaven the gods became afraid of the water-spout." Thus all through legendic lore, we can trace the belief of man that the deluge was a mighty down-rush of waters from the celestial abyss, and that Bel, the sun-god, was the author of it all. Deucalion's flood, which is evidently the Chaldean modified to suit the Greeks, sets forth many of the same occurrences and features just named, and I need not add them here.

Among the Hindoo legends I find the following: One morning water for washing was brought to Manu, and when he had washed himself, a fish remained in his hands, and addressed him thus, . . . "a deluge will sweep all creatures away . . . the very year I shall have reached my full growth the deluge will happen. Then build a vessel and worship me. When the waters rise, enter the vessel and I will save thee." There is also another form translated from Hindoo, which is evidently taken from the Chaldean or Biblical. "In seven days the three worlds shall be submerged." Among the Iranians, the sacred books relate how the original ancestor, under the name of Yima, is ordered to construct an enclosure and cause men and animals to enter it in order to escape destruction from a deluge. Now all these legends, pointing to the general destruction of man and beast, and the construction of some ship, chest or enclosure to preserve them, leads directly to the annular system as the source of the destroying waves. Both the legend of Ogyge's flood and that of Deucalion, refer to the offices of the ark in the preservation of a few persons, from a general destruction. The Koran says, "All men were drowned save Noah and his family; and then God said, 'Oh Earth, swallow up thy waters, and thou Oh Heaven, withhold thy rain,' and immediately the waters abated." The Egyptians seem to have had a correct knowledge of the deluge, for they told Solon that there had been many;

and also told him of " one great deluge of all."—Plato's " Dialogues."

My readers no doubt are familiar with the fable of Phæton, son of Helios, who harnessed the steeds of the sun to his father's chariot, but because he was not able to keep them in the path of his father, produced a general conflagration and destruction. This evidently is a myth which arose from the fact that, as the upper vapors declined; the sun came more vividly into view, and as the waters of the deluge fell, it became visible in all its might and majesty, and all terrestrial nature, including man and beast, all vegetation unaccustomed to the blaze of the sun, must have suffered from its direct action. So the sun-power would be remembered with the same vividness as the deluge, by those in lands not so greatly affected by the fall of waters. And in the land of Egypt, above all others, I presume this phenomenon would be more thoroughly remembered. Now it is not a little remarkable that the Egyptian priests should tell Solon that the fable " Really means a decline of the bodies moving around the earth, and in the heavens." The declaration could not be plainer if it said, it " Really means the fall of revolving vapors from the heavens." (See Plato's " Dialogues.") This volume might be filled with such traditions, from almost every nation, kindred and tongue on the earth. Hindoos, Brahmins, Chinese, Sandwichers, Fijis, Peruvians, Mexicans and Alaskans have all preserved deathless memorials of this great event.

The classical student is now prepared perhaps to see that the rock of the annular theory underlies the entire system of Eastern and Western mythologies;—

that the light radiating from the former, illumines and simplifies the latter to a marvelous extent. Let us admit the interdiluvian world to have been in its earlier periods the scene of perpetual day lighted up by a yet invisible sun, and a hundred mysteries are readily solved. First we may readily understand why the sacred historian informed his readers that day and night should forever alternate after the deluge when the sun came distinctly into view. (See Gen. 8: 22.) Then we may readily see the sun-power represented by Osiris, Hercules, and the Apollos to have been looked upon as the forming, conquering, and renovating deities of the ancient world. The absence of the sun's direct light and actinic energy having formed the Edenic clime that characterized the world in which the infant human race was nursed, it must be plain that as the vapors grew thinner in the equatorial regions, as they spread toward the poles of the earth, the sun came gradually into view, and its coming was the physical agent in the Creator's hands in putting an end to the Eden climate. The sun's absence made Eden, and therefore his presence destroyed it and drove man from his genial home. We are impelled to this conclusion by inexorable law. The solar orb then coming into view more fully as the vapors thinned away, and the climate of the Eden world growing colder necessarily at the same time—a physical curse thus coming upon the earth—mankind must have looked upon the sun as the agent in the hands of the supreme Arbiter, to punish them for their sins. In short, that coming conqueror was looked upon as the cause of the deluge. With this, as the agent of the Omnipotent Hand, in the estimation of man for punishing the wicked, not only is the darkness of mythology largely expelled, but the whole Edenic history is wonderfully illuminated.

Then when I read the remarkable passages of Job 38: 12, 13, " Hast thou commanded the morning, since thy days; and caused the day-spring (sun) to know his place; that it might take hold of the ends of the earth, that the wicked might be shaken out of it?" it shows plainly to my mind that the famous patriarch lived in an age when men believed the sun came in to punish the wicked! This, and several other passages, I cannot now enlarge upon, leave not a doubt upon my mind, that much of the book of Job was penned in interdiluvian * times. We are now merely approaching some of the momentous conclusions to which the annular theory must inevitably lead. Conclusions that must make sweeping and radical changes in opinions, as the theory gains a place in the philosophic credence of men.

* I use the term interdiluvian to represent that period extending from the expulsion of man from Eden, to the final fall of the waters upon the earth. As it is evident man was deprived of his Eden clime by declining vapors, first in the polar regions, this period must have been one of great climatical changes; in fact, must have been an interdiluvian one.

CHAPTER VIII.

OCEANIC DOWNFALLS AND AUGMENTATION OF TERRESTRIAL WATERS.

A CONSIDERATION OF THE EVIDENCE IN SUPPORT OF THE CLAIM THAT THE WATERS OF THE OCEAN HAVE BEEN GREATLY INCREASED IN VOLUME IN VERY RECENT GEOLOGIC TIME.

The reader must now see there is no possible escape from the conclusion that if the Noachian waters fell to the earth, as our theory predicates, there is now a greater volume of ocean on the earth than before the deluge. That the shores of the ocean are further inland; that the estuaries and bays, straits and seas communicating with the ocean are wider and deeper; and further, that this must be the condition of the world at large, except where recent elevation has taken place. If, then, it can be shown that this condition of the oceanic world does obtain, it will be taken as another link of valuable evidence. Again, when this evidence is confronted by that now universally held, that all the submerged regions of the earth merely sank and let in the waters of the ocean upon them; if the unphilosophic nature of the claim can be shown, the evidence becomes positive and conclusive; for the submerged tracts of the earth either sank or the oceans have been augmented.

It is true that many portions have sunk; that some parts of the earth are sinking to-day, and others steadily rising. This phenomenon must needs be, while rivers bear their measureless burden to the sea; while currents transport solid matter from one part of the earth to another, and the conservation of energy remains a fixed law in the universe. It is calculated that the solid matter annually carried down by the great Mississippi is sufficient to cover 640 acres to the depth of 240 feet. This floated matter is deposited in the Mexican Gulf, or carried by currents into the adjacent oceans; and we must not forget that every pound of matter thus transferred is an energy transferred! In the course of one thousand years, one thousand square miles of oceanic bottom would be covered 240 feet, by an actual accumulation, and the underlying beds would support a pressure measured by millions of tons. This enormous pressure upon the underlying rocks is so much transferred energy converted into mechanical heat. This mechanical heat must of course, expand the rocks thus under increased pressure; and as there is apparently no measure of this expansive force, what rock can there be that will not yield to the force exerted? At the mouth of the Mississippi the continued additions of mechanical pressure by the constant deposits, borne down by the river, give rise to bubbling and steaming hillocks of mud,—veritable miniature volcanoes. As increase of pressure must give rise to increase of heat, unit for unit, it is plain that excessive increase of pressure must produce corresponding increase of heat. Is it not a legitimate conclusion, then, that if all the sediment and precipitated matter carried into the Gulf of Mexico were deposited on its bottom, and not borne to a great extent into the Atlantic, its coast would be like that of the

Mediterranean Sea, lined with numerous mountains and the scene of active volcanoes?

The great Mediterranean is certainly a grand example of the conservation and transfer of energy. Many large rivers pour into it from all sides, bearing such enormous volumes of sediment that is not carried to the ocean, but is constantly settling upon its bottom; and the frequent and appalling eruptions so well known in modern times, cannot but be pure results thereof? Can scientists find any other vents than volcanoes and earthquakal agitations for this force employed? It must be accounted for. It cannot be lost. And the question might well be asked: Can volcanic eruptions have any other cause than that of transmitted or transferred energy? As we look around the globe and see all its volcanoes located in regions where transported sediment is accumulating—i.e., in and around the ocean borders, and see that no volcanoes are located where no sediment can accumulate, can we for one moment doubt that we have here the true cause of volcanic eruptions. As the underlying rocks expand by the increase of heat, arising from additional sediments continually gathering upon them in the seas, they must fracture and crush into neighboring rocks; which crushing must give rise to centers of fire susceptible of fusing the beds around them. And it is conceivable that sufficient sediment may gather over a bed of rock to liquefy the latter. About 65,000 feet of steel blocks piled one upon another will give rise to sufficient heat to melt the lowest blocks, or at least to render them plastic. Hence, the reasonable conclusion that the lava that issues from a volcano is the deep bed-rock fused by pressure, produced by lateral expansion. Thus we may behold even here the grand effort of solar action. Solar heat raises the vapor on high; it falls as rain, on hill and plain, swells into a stream, or feeds a fountain, and gathers sediment as it runs through its channel to the sea, where it adds its increment of mechanical heat to fuse the rock. So that the force employed in the grandest volcanic eruption is the same in amount as that employed by the sun-beam in raising that vapor from the sea to the clouds.

This little digression will prepare the reader to understand that as sediment is continually accumulating in some regions, and being removed from others, there must be rock expansion going on continually in some regions, and continual contraction in others. Expansion must elevate the earth's crust. Contraction must lower it. The reader must see that this is law, and must also see herein an adequate cause for the sinking of some coast regions, and the elevation of others. It is an extremely slow motion, which, arguing an extremely slow accumulation or diminution of mechanical energy would seem to point alone to the cause here supposed.

But while in the neighborhood of the ocean's boundaries such oscillations may occur, we surely could not expect such to obtain to any extent among islands in mid-ocean, or on coasts where for many thousand miles no rivers of importance exists. The elevation and submergence of such coasts must be attributed to other causes. Now the continents have all been lifted from the oceans, and can it be possible that they could be raised to their present position by any other than a solid bed of intruded or expanded matter. They were lifted by a force directed from the oceans, as all will admit. Did the slow accumulations of sediment accomplish this grand result? If so, why was that energy put to work at successive periods, and attended with sudden and abrupt changes, and extermination of specific forms? The same force at work to raise the coast of Norway could not lift a continent and put eternal props under its adamantine sills. Wherein, then, can we find a competent cause? Can the annular theory supply it?

Let us suppose a downfall of water at this age should raise the surface of the ocean 50 feet above its present level. The reader will see that every ton of water thus added to the pressure on the ocean's bed must be converted into so many units of mechanical heat in the granite foundation of the aqueous beds, causing an expansion which nothing could resist, and directing this measureless force towards the continents. The only result that could take place is evidently the forcing of matter, solid or plastic, from pressure under them, ending in their elevation; or the plication of their margins into mountain-folds. Now such things have taken place again and again in the past ages of this planet. Grand convulsions, coupled with universal oceanic baptisms, and change in life-organisms, have repeatedly taken place; and the up-lift always directed from the ocean. As an installment of annular matter is necessary for the baptism; necessary for the transfer of a competent energy; necessary for general extermination of species, how can we avoid the conclusions that the oceans have many times by immeasurable additions climbed upon the shores of the world?

It is clear, then, that the waters of the deluge referred to in a former chapter, if anything nearly so tremendons as claimed, must have resulted in crust-folding, or elevation, especially in the neighborhoods where great river systems carried their detritus into the seas. It surely is not necessary for me to enumerate and specialize the localities of the earth, well known to the geologist, that have in the most recent geologic times been lifted from the ocean's wave, with the shells of existing species. The New England coast was elevated, as all know, many feet since the last advance of glaciers there. This being the case, we must look around us for the rivers that bear their burdens of continental detritus to the seas. Well, the great St. Lawrence washes the feet of New England on the north, in a valley so new that thousands of rock-bound islands gem its waters. So new, I say, that it has not yet swept its channel clean, and must therefore have been recently elevated, with its surroundings, from the sea. It is one of the mightiest excavations of the earth. One can not look upon its wide reach of flood-ground, and lofty facades, and not ask, what has became of all the matter borne from this valley? One comprehensive glance at the great banks of Newfoundland will answer the question. And now when we see the thousands of square miles of detrital offspring of the Hudson's waters, how can we conclude otherwise than that those great beds were the products of diluvial times? In short, does not everything seem to argue the transmission of a competent energy by an adequate cause, by means of which New England's recent elevation was effected? As this phase of mountain-making will be fully treated elsewhere we will now turn our attention to the more direct question of oceanic augmentation in modern times.

Suppose the reader could see at one view all the river estuaries of the earth! Knowing their delta approaches to have been built up by enormous accumulations of detritus, forming in many instances, as in the cases of the Mississippi, the Nile and the Ganges, beds several hundreds of feet thick; and knowing, too, that the same sedimentary deposits are being made at the outlets of rivers, emptying into inland lakes, we can readily understand that there must be a difference in the appearance and character of lacustrine deltas and estuaries, when compared with those of the oceanic borders. Let us examine the great Lakes of North America, emptying through the St. Lawrence into the ocean. It is readily seen that the waters in these lakes cannot rise permanently while their present channels of outlet remain; that however abundant the additions to their waters, so long as river erosion continues, these lakes must grow more shallow with the flow of centuries; that in a few thousand years cities now planted upon their shores must become inland towns, unless they follow the receding waters. This recession of lake waters, and falling of the lake level in the region referred to, have been going on for unknown time. But it must be seen that as the waters fall, the pitch or decline of the river current, as it enters the lake, must also increase and follow them; so that in course of time the river's course through a lacustrine delta, would be between walls continually increasing in height. How emphatically true this is of the lakes of all the earth, I need not say. It is simply a matter of observation, which every one can verify. And the question here maintained is then apparent:—the recession of waters from an estuary deepens the channel and increases the pitch and flow of a river current.

The high sand banks contiguous to the lake estuaries of this country, which many of my readers must have noticed, simply prove that the lake level has fallen. Then, again, we must see that if the level of the lake waters should be elevated, the results would be just the reverse. The pitch of estuary currents would decrease; the mural escarpments of river courses would decrease in height and the delta formation become one level expanse of detrital accumulations. We may now apply a decisive test to the river deltas on the oceanic borders of the world. In the search of twenty years I have been unable to find an oceanic delta, with its accompanying estuary, that possesses the lacustrine characteristic of increasing pitch. Where the land has been elevated so as to show walls and deep delta channels at the mouths of rivers there are features which the keen eye of the geologist may readily see, as strikingly different from those in lake regions. The elevation of a river mouth throws a volume of water back upon itself, and the deposit is no longer the same as before. All the great rivers of the earth present, however, the very same appearance, we would find, as stated above, in a lake delta and estuary, whose water level was elevated by an increase of the volume of water. It is the rarest circumstance to find a river of much importance flowing with a rapid rush of waters into the sea. There is a wide expanse of land scarcely above the sea level, washed daily by the tides, and the river flows lazily along continually dropping its load of sediment. No river bluffs. Almost every sign of channel or river escarpments has been obliterated, if such ever existed. Why is this so universally the case? It simply proves that land elevation at the mouths of rivers is exceedingly rare. It proves that while the accumulations of sediment, or delta deposits, are continually pushing the wave oceanwards, by the growth of land area, the level of the oceanic waters has been elevated in modern geologic times, so as to obliterate channels and walls that must inevitably grow in the lapse of ages, at river mouths. For, with a call ceaseless as the flow of time, and with an appetite as insatiate as death, the hungry earth is absorbing its waters.

The problem, then, is reduced to this: Either the oceanic waters have been augmented in volume, by additions thereto, in modern geologic times; or, the land at the outlets of almost all the rivers of the globe has been sunk. Which of these is the more probable? Nay, can it be possible, that so nearly all the oceanic deltas of the earth could present the actual appearance of an increase of oceanic waters, unless such an increase had taken place? If these deltas, even to a limited extent, exhibited the inclination due to local elevation, we might calculate between the probabilities of land depression and emergence; but when it is all depression, there can be no probabilities, and certainties only come within the purview of our calculations.

Again, look at the great Pacific, studded with island gems, that are, as is well known, the summits of mountains submerged. Here are millions of square miles of submerged lands, as proven by coraline formations,*

* "A Melbourne journal describes a remarkable piece of coral taken from the submarine cable near Port Darwin. It is of the ordinary species, about five inches in height, six inches in diameter at the top, and about two inches at the base. It is perfectly

that have in modern geologic times succumbed to oceanic inroads (i.e., oceanic elevation), and which would to-day be a grand continent peopled by industrious millions, and covered by luxuriant tropical vegetation, if its waters could be lowered to the extent of 80 or 100 fathoms. Then when we turn to the eastern coast of North America, we find a vast region of coast-line just submerged, and glean from the "Coast Survey" the remarkable fact that from Nova Scotia to Florida, and from Florida around the whole boundary of the Mexican Gulf, there are the submerged shore-lines of a former continent, far out from the present shore. That for nearly the entire extent of many thousands of miles of coast, these shallow waters of about 30 or 40 fathoms deep, roll shoreward from the mighty depth of the ocean. Beyond this actually known and surveyed ancient shore-line, now from 80 to 100 fathoms under water, the lead and line plunges suddenly to a depth of 200, 400, 1,000 or 1,500 fathoms. From the British coast-survey we learn that the British Islands are surrounded by the same character of coasts, a mere playground for waves. Gradually the waters deepen from the present shore ocean-ward, until we suddenly arrive at the old coast-line beyond which lies the abyssal deep.* The German ocean and the Norwegian waters are so shallow that if they were lowered 30 or

formed, and the base bears the distinct impress of the cable, and a few fibers of the coil rope used as a sheath for the telegraphic wires still adhering to it. As the cable has been laid only four years it is evident that this specimen must have grown to its present height in that time, which seems to prove that the growth of coral is much more rapid than has been supposed."

* For the first 100 miles out from New Jersey the ocean deepens only three feet a mile, or 300 feet in all, while 18 miles farther out the water is 6,000 feet deep, and 250 miles out is 2½ miles deep.

40 fathoms they would expose a vast stretch of level continent on the northwest of Europe. The whole coast of northern Europe and Asia presents the same characteristics. When we turn to the waters washing the eastern shores of Asia we find the same; and wherever the southern shore has been surveyed from 60 to 100 miles from land, throughout the whole coast from Java to the Gulf of Aden, we find the same shallow oceanic water; and beyond its boundary, the deep ocean. Turning again to North America, from Columbia River to Behring's Strait we find shallow ocean. From the Columbia River southward to Cape Horn, I have but little information respecting the character of the sea bottom; but from many places come authentic information that the Pacific waters now roll over submerged forests near the shores. The same kind of reports are heard from the Atlantic coasts of South America and Africa.

But most fortunately we have information that none will dispute from the very midst of the Atlantic Ocean, many portions of which have been surveyed and mapped. The U. S. sloop Gettysburg several years ago, when about 300 miles west of Gibraltar, anchored where the sounding line revealed a depth of only 32 fathoms. The British ship Challenger and the U. S. ship Dolphin have traced the course of a submerged continent in mid-ocean, and seem to have demonstrated the former existence of a long insular continent, nearly mid-way between the two existing continents, and running nearly parallel with the general trend of the Atlantic Ocean. The character of the deeply cut and channeled bed of these mid-ocean ridges shows that they were in recent geologic times subject to aerial denudation.

Thus it seems there is an abundant evidence, so far as we are able to glean from the physical testimony of the universal oceans, that its waters, the world over, stand higher to-day upon the shores of the continents than they formerly did,—than they did in recent geological times! How can we conclude otherwise? Can it be possible that during the same age the great Pacific continent covering millions of square miles was submerged; the whole Atlantic bed; the Indian Ocean, and the North Polar seas, should all climb from 30 to 40 fathoms upon the shores, because of a subsidence of the land alone? When Behring's Strait was made to connect the polar waters with the Pacific; when the Strait of Dover separated England from the continent of Europe; when the Strait of Gibraltar connected the Atlantic with the Mediterranean, the oceans either increased in volume or the continents sank. And when we know that every drop of the immense oceans that now wash the shores of the world, has actually fallen to the earth's surface since its igneous era closed, and since the very pointing of eternal law, from whatever field we may view them, shows that these oceans may have fallen in terrific and overwhelming cataclysms through the measureless lapse of ages, and not all in primitive times, why should we be slow to admit the grand and philosophic thought, that recent geologic times closed with a vast augmentation of the waters of the earth? Most impressively is this consideration forced upon us, as we turn to the records of the glaciated continents, and reflect upon the immensity of the snow-fields that filled the valleys of almost every land, till the face of the planet gleamed in universal ice. And when we take one step further in the investigation, and find that the closing of the glacial period was the commencement of modern oceanic inroads; that the oceans climbed up the shores as the glaciers melted away; and further find that terrific deluges of water were urged for unknown time down the innumerable valleys of the earth, there seems to be no foothold for skepticism on this point.

I will now collate some interesting facts relative to this phase of the annular theory, some of which have been presented by well-known authors and scholars. (From Geikie's "Ice Age," page 91.) "From these and similar facts geologists have been inclined to infer that at the time the *mer de glace* covered Scotland the whole of our country (Britain) stood at a higher level relative to the sea than now; in other words, that a large part of what in these days forms the floor of the sea was at that time in the condition of dry land."

Again (same page), "The German Ocean between England and the coast of France, and the Netherlands, does not average more than some 150 or 160 feet in depth; and the soundings show that the water deepens very gradually northwards."

And while we are considering this part of the geological field, we will examine some further evidence of oceanic elevation. Submerged peat-beds containing trees and trunks of oak, pine, hickory, walnut, etc., are witnesses of recent advance of oceanic waters. These are found in abundance around the coasts of England, Scotland and Ireland, not only in connection with the main land, but in the small outlying islands of the British seas. (Geikie, pp. 294 and 295.) (See also Sinclair's Acct., vol. xvi, p. 556.)

On the Frith of Tay are larger tracts of submerged

peat-moss, containing hazel-nut and alder, many feet below full tide. (Edin. Rog. Soc. Trans., vol. ix, p. 419.) The same are found on the shores of Tiree and Coll. (Edin. Jour. Phil., vol. vii, p. 125. They are found in abundance on the northern coast of the continent, from France to Denmark and Sweden. Submerged forests abound along the coasts of Brittany, Normandy, and the Channel Islands, as well as off the shores of Holland, and also on the Alaskan and Siberian shores.

Can we come to any other reasonable conclusion than that the northeast Atlantic and the German Ocean have largely augmented their domain, in comparatively recent times? We have reports of submerged forests on the wide circuit of the ocean world. Scarcely any considerable part of the globe whose boundaries lie by sea, does not exhibit some such evidence.

Captain Herandeen, who spent many years on the Pacific Ocean, has given some interesting and valuable evidence in regard to the great insular continent that now sleeps under its waters. I draw briefly from one of his narratives: "But there is other evidence which is more interesting, because it relates to the great decay of a great race of people that once inhabited the region. A few years ago I stopped at Pouynipete Island, in the Pacific, in east longitude 158° 22′ and north latitude 60° 50′. The island is surrounded by a reef, with a broad ship channel between it and the island.

"At places in the reef there were natural breaks that served as entrances to the harbors. In these ship-channels there were a number of islands, many of which were surrounded by a wall of stone five or six feet high, and on those islands there stood a great many low houses built of the same kind of stone as the walls about them. These structures seem to have been used as temples and forts. The singular feature of these islands is that the walls are a foot or more below the water. When they were built they were evidently above the water and connected with the main-land, but they have gradually sunk until the sea has risen a foot or more around them. The natives on the islands do not know when these works were built; it is so far back in the past that they have even no tradition of the structures. Yet the works show great signs of skill, and certainly prove that whoever built them knew thoroughly how to transport and lift heavy blocks of stone. Up in the mountains of the island there is a quarry of the same kind of stone that was used in building the wall about the islands, and in that quarry to-day there are great blocks of stone that have been hewn out ready for transportation. The natives are in greater ignorance of the phenomena that are going on about them than the white man who touches on their island for a few hours for water. There is no doubt in my mind that the island was once inhabited by an intelligent race of people, who built the temples and forts of heavy masonry on the high bluffs of the shore of the island, and that as the land gradually subsided these bluffs became islands. They stand to-day with a solid wall of stone around them, partly submerged in water."

Thus we not only meet with the strongest evidence that the waters have arisen on the shores of the continents, severing in numberless instances islands from the main land, as England from Europe, and the West Indies from the American continent, but in the very heart of the oceans we find the same testimony in imperishable monuments. How often has this aged world of ours been shaken by the mightiest revulsions! How many races similar to man, the masterpiece of the Creator, have felt the blow of inexorable fate!—races of sentient beings that may have lived amid the flourish of empires, and the shock of death, till swept as by a stroke from the earth, before the Adamite or Edenic man came upon the scene. Here, blocks hewn from the mountain quarry have significant meaning. Left in confusion they argue that the workmen were suddenly driven from the quarry, just as in many other cases in other parts of the earth. The ancient copper mines of Lake Superior, in several instances, show that the old miners that used the flint and other stone implements, were suddenly swept from their place of work, never to return. Their axes, hammers, wedges, etc., left lying around in the utmost confusion, and covered with flood-detritus in deep excavation, tell an unmistakable tale of sudden and violent catastrophe.

All over the ocean world then—where rivers empty their waters; where inlets lie embosomed in forest and rock; where straits separate mighty continents, and connect ocean with ocean, and sea with sea; where islands rise from the restless wave, in the very midst of boundless oceans; wherever we may chance to turn our gaze upon the watery world,—we see, it seems to me, the most surprising evidence that the earth's waters have been greatly augmented in modern times. Tell me, what else could have raised the waters so generally over the earth? Is it not plain that the vast expanse of the Pacific continent subsiding, would have drawn such vast volumes of water from the shores of other continents, that rivers would be free to pour their waters with rapid flow into the sea? Dana says the sunken continent of the Pacific is 6,000 miles long, and from 1,200 to 2,000 miles broad, and makes out that it has sunk more than 3,000 feet. If we make this less by 100 times, we must even then, by some means, find a source for those waters which again filled the estuaries after they were drained by the ocean's sinking bed. We cannot conceive of such an enormous area of sea bottom, sinking even to the extent of a single foot, without increasing the rapidity of the flow of river waters at their outlets. But where, on the wide face of the earth, do we see this to be the case? Since we see the reverse to be almost the universal rule, it seems to me we are simply compelled to admit that the ocean's waters have climbed upon the shores of all the continents. Now the fact that such coasts as those of Norway and Sweden and some islands in the North Pacific have been elevated in modern times does not in the least oppose these ideas; for the fact that we are able to prove that they have been elevated from the sea, only proves that they, too, were submerged, increasing the necessity of admitting the fact of oceanic augmentation.

What, then, does that buried continent prove? Does it not prove that a mighty deluge did desolate the earth? And as it is a fact which every one is forced to admit, that in the Noachian period a vast deluge of waters did come from beyond the region of clouds and rains, we cannot avoid the conclusion that the ocean's volume was then increased.

Now let me ask the reader, what conclusion must we draw from the array of facts now before us? Did the primitive vapors return to the earth as they condensed in primitive times, contrary to law universal and unchangeable? Is it not within the conception of every one, that if all the waters of the earth fell on the archæan sphere, then there never was a deluge? That there never were waters above the firmament? That the sun came into view in primitive times? That consequently there never was a day of physical rest; nor a day when it did not rain, nor a time when man dwelt naked on earth; nor an Eden clime? Then the rainbow was a common occurrence in all times, and can in no sense be a token of God's promise to man. Then man always lived in this present environment, and his days were always three score and ten years. In short, the whole Edenic narrative becomes one meaningless tissue of contradictions, beyond the pale of law. If the waters of the earth were not increased under the cognizance of the human race, what can the first eight chapters of Genesis mean? Refuse to admit this philosophic necessity, and we are plunged into the darkness of midnight. Ineffable harmony and beauty becomes hideous disorder and deformity. And now when we take a comprehensive glance at the seas of the earth, and can find but one grand chain of evidence in support of "upper waters;" in short, as we find the globe today one marvelous and comprehensive argument, in defense of Edenic history,—an argument which is the voice of law; I must say, with emphasis, the earth's annular system was a physical and necessary fact.

We started on our tour of investigation with the infant earth wrapped in the swaddling garments of flame, and rocked in the cradle of primitive fire. Measureless cycles rolled away, and then we saw the youthful orb flying through space, a glowing and vitalizing sun. Revolving around the eternal throne of implacable law, as its fires smouldered away its oceans gathered around it. Away down the vistas of time we see plans perfected. The world unfolds at the beck of Deity. Man, the masterpiece of the Omnipotent Designer, familiar by actual contact and knowledge with the great canopy of vapors, has sent down to us a most faithful and inexpressibly harmonious history. The rock-bound records confirm its details. The ocean unites with the inevitable verdict, and the annular theory stands a citadel of rock. We have proven it first by mathematical reasoning and philosophic necessity. Then we have proven it by the mineral character and philosophic disposition of strata. And again we have proven it by analogous facts relating to our sister worlds, belted and ringed under the reign of law. Then, again, we have proven it by the action of our own satellite. Then we have taken the records of man, rude and mysterious, and have shown by the very nature of those ancient writings that they declare and redeclare, again and again, the truth of my claim. So that if all other evidence were cast aside, if all the demonstrations, and doubly riveted links of testimony before adduced, were entirely left out of the argument, the first eight chapters of Genesis alone afford a proof so abundant and positive that no sane man, it seems to me, can for a moment doubt that they are a true and faithful delineation of the earth's annular appendage.

What kind of a chain of evidence have we then, with all these witnesses testifying to the same thing? And after we have so firmly established this thing; when we examine the waters on the earth and find that they bear witness to the same thing in such a way as to become a demonstration in themselves, surely my readers will pardon the egotism: I have proven so far as positive evidence can prove anything, that this earth had an annular appendage from the remotest period of archæan time, through the ages to the days of Noah.

Now if the reader choose he may cast all this evidence aside, and we will begin a new series of demonstrations. He may throw away every page of testimony I have given, and I will prove the same great truth to him by testimony from other fields of investigation. We have scarcely entered the field. Our work has just begun.

CHAPTER IX.

SOME TOPOGRAPHICAL FEATURES THAT PROVE THE DECLENSION OF EXTERIOR MATTER.

It must be admitted by all who concede the truth of the nebular hypothesis, by all who admit that this planet was ever in a state of igneous fusion, that the mass composing the great ocean of primitive vapors that surrounded it was impregnated with vast quantities of elementary mineral and metallic matter. This is so evident that I need do nothing more now than call the reader's attention to it. I also need but call his attention to the fact that when the upper waters, or vapors, with their associated matter, fell to the earth, they must have made temporary seas, lakes and ponds, etc., in all parts of the earth where they fell. The narrow channels of thousands of rivers could not permit the mighty floods to immediately retire. In those lakes and seas would be deposited the precipitates and exotic solid matter of the annular waters, and especially so in regions beyond the tropics; and the nearer we approach the polar regions, the more abundantly we would find this exotic matter. This must be essentinally the case, if there be a polarwise tendency to declining belts, etc. But what kind of precipitates must we expect to find? Let us determine this matter before we search for it. First and most important of all the elements of the earth's crust is carbon. Of the sixty thousand feet of aqueous beds there is probably none of which this element does not form an important constituent. Hence we have no possible means of escaping the conclusion that the earth's primitive atmosphere,—largely the products of igneous action,—contained vast quantities of carbon sublimed or distilled in the earth's glowing crucible. Let the reader see, before he proceeds farther, that we are irretrievably committed to this conclusion. As it would be an utter impossibility for this earth to be now reduced to a molten condition without sending up an immensity of unconsumed carbon, in the form of smoke, so it must be a settled and absolute fact that the primitive burning earth, from the very day it became the seat of fiery fusion, repelled from its heated bosom, and held in suspension, unconsumed carbon or smoke. Every chemist familiar even with the rudiments of his science, will tell us this must have been the case. To conclude otherwise would force the admission that the primitive atmosphere was an ocean of oxygen, which simply could not have been the case. Hence we are driven to this unavoidable end.* The primitive earth was a burning world, and therefore a smoking world, and that unconsumed carbon commingled with the annular vapors just as it would to-day, in the form of black, sooty, pitchy matter. As we cannot avoid the conclusion that unconsumed carbon was mechanically combined with the upper vapors, so also we are made to admit that it mingled with them in the form of soot. Can the reader find a flaw in these statements? But if this sooty, carbonaceous matter mingled with the exterior

* All the matter composing the earth fell to it, either before aqueous attrition began, or afterwards, or partly before and partly afterwards. Then all the matter composing the aqueous crust fell to the earth in the later stages of its evolution. What reason, then, can be urged against the fall of the tellurio-cosmic matter being continued all through the geologic ages, at the same time that aqueous denudation went on?

vapors, then they fell in company. And the waters that stood in "seas," "lakes," "ponds," etc., at the time of the deluge, deposited this carbon as a layer of black carbonaceous mud upon their bottoms; for we cannot admit that even the last remnants of the annular waters were not associated therewith, just as the belts of Jupiter and Saturn are darkened by such sooty matter to-day. Now we may see some meaning in some of the flood legends, which declare that the waters of the deluge were a "pitchy blackness."

If, then, we succeed in finding this black carbonaecons matter at the bottom of inland seas, lakes, etc., or spread out over extensive plains, that were formerly covered by standing water, we must see a wonderful dovetailing of facts, that add strength to our theory. I hold such deposits must be found in order that the theory be fully vindicated. I have no need to tell geologists of the tens of thousands of lakes, planted in the drift deposits of North America and Northern Europe, whose bottoms are known to consist largely of the very carbon we need to find to sustain our views. Hundreds of them have been drained in Northern Ohio and in Michigan and other States, some for agricultural purposes, some in the construction of canals and railroads, and almost invariably they present the same features. Many of these ponds that I have personally examined had no vegetation, and therefore the carbon could not have been a peat formation. While those which had been converted into swamps, and covered with peat-growth, had the peat formation underlain by the primitive carbon which everywhere presents its own characteristics. These things are subjects of ocular demonstration, which any one can verify for himself. There are more than ten thousand ponds and lakelets in Minnesota alone, and so far as I have been able to learn from them, they abundantly support the claim here made. They are found in many parts of northern Indiana, Illinois and Iowa, where I have personally examined some of them, and find the same evidence. A layer of black carbonaceous mud lies at the bottom of the lakes that have been thus far explored,—carbon that cannot be called peat! and since there can be but one other source, its origin is apparent.

I suppose there are but few of my readers who are not aware of the fact that a black carbonaceous soil is the superficial covering of many of the northern and northwestern States,—a coating of exceedingly black, soot-like matter, strikingly different from that of the adjacent States. Now since it is well-known to geologists that all this region thus overlain was once the bed of a vast inland sea, covering more than half a million square miles; in the eyes of the geologists at least, we have one feature established that points to a deposit of light, primitive carbon from on high—viz., the fact that a sea existed, which was necessary for its distribution and deposition. But as these pages are intended for all readers, my next duty evidently is to prove that such a sea did exist, and then to prove that the superficial covering is a deposit of annular soot.

Again, let us see that we start with known and universally admitted premises. On the west of this great basin rises the mighty wall of the Rocky Mountains, and since the close of the tertiary age it has been a great divide between the waters running westward and those running eastward. Between the waters of the Arkansas and those of the Missouri, is another divide running eastward from the Rocky Mountains through southern Kansas, and abruptly terminating at the Mississippi River, as the spurs of the Ozark Mountains. This Ozark range is another wall vastly older than either the Allegheny or the Rocky Mountains. The archæan beds that compose much of its course prove that it was one of the oldest wrinkles on the continent.

With the exception of the gap through which the Mississippi flows this ancient wall is continued unbroken till it joins with the mountains of Tennessee and Kentucky. Here, then, we have a western wall dating back to the tertiary, and a southern wall, much older, broken only by the waters of the Mississippi. Now from a point a few miles south of the mouth of the Kaskaskia River, the Mississippi River runs between walls more than 700 feet higher than the bed of the stream. A wide channel has been cut through this southern wall in modern geologic times. For there is the gap through which the waters now run; and there is the ancient wall continued on either side of the stream. Suppose, then, this great gap were again filled up; any one can see that it would dam up the waters which would again arise and submerge much of the Western States, and cause the waters to run through the only other outlet possible—the St. Lawrence valley,—thus forming a great inland sea, the very object we desire. Thus when the Ozark were upheaved among the oldest plications of the earth, the new-born continent, from about the 35th or 36th degree of north latitude, drained its waters northward, and those from the Rocky Mountains afterwards ran eastward. But the great Canadian highland, separating the waters of the St. Lawrence and the Great Lakes from those emptying into Hudson Bay, is considered by all geologists as the oldest range of highlands on the earth. Here, then, we have a north wall bordering the Great Lake or inland sea-basin, reaching from the Atlantic Ocean to the Rocky Mountains, and interrupted only by the elevated depression of the Red River Valley of the North. As these are simple facts which all geologists will admit, I need not advance any further evidence to prove that the Great Basin drained by the North Mississippi and its tributaries is bounded on the north, west and south by walls of great age. But if this be true, we must admit that immediately following the tertiary age all the waters of the North Mississippi Valley, and those of the Ohio, flowed eastward and northward, and emptied their waters through the St. Lawrence into the Atlantic; for it is a matter of universal consent that the New England mountains are geologically of very recent origin. Hence there was a time when there was a vast basin, walled on all its sides, except the eastern,—an age when New England was covered by the sea,—and a vast river running from the Rocky Mountains received its hundreds of tributaries from all sides and emptied its waters as the St. Lawrence now does. One glance at the ancient rim of this basin must force this conclusion. What a wonderful revolution must have taken place in the drainage lines of the continent! To conceive the great Missouri, threading its way among mighty forests across the States of Iowa and Illinois, and emptying into Lake Michigan, may seem to border on the visionary. But let us remember that grander and mightier changes have left their way-marks upon the earth. It is the only conclusion we can come to, as we reflect, that two great parallel primitive mountain ranges—the Ozarks and the Laurentian Ridge—extended east and west across the infant continent, when the Cordilleras were heaved from the deep. But lest the reader may think I have strained the evidence here produced, I will compromise so far as to only claim a probability that this was the drainage system of this Great Basin, and we will bring in other testimony to establish this point.

If it be true that this was the condition of the basin at the time referred to, then, when the New England Mountains were lifted from the ocean, it threw a great barrier across the St. Lawrence, and forced its waters back upon the valley commingling the marine faunæ with those of fresh water. How truly this is the case, all geologists familiar with this territory know full well. Imagine then a new wall raised upon the eastern shore of the basin. Inch by inch the confined waters accumulate. The St. Lawrence Valley becomes first the bed of a salt water lake. As the waters increase it grows brackish, and finally fresh. The location of marine faunæ in abundance in the country east of Montreal, and fresh water shells on the west, and the commingling of them in the elevated terraces near Quebec, certainly strengthen the claim I have advanced. But when we behold the wonderful mural heights a few miles below Quebec, between which the St. Lawrence now flows, a still stronger evidence is added. How did this river ever force its way through this embrasure? On either side of the river are mountain heights that doubtless were once joined as a natural breast-work across the stream. Geologists will all admit that this eastern wall must have been lifted more than 500 feet above the level of the ocean, in very recent times.

But this much of a wall across the St. Lawrence would have backed its waters, and have buried Lake Ontario more than 300 feet. Lake Erie would have spread its waters into Lake Michigan, and all northern Illinois and Indiana, and much of Iowa would have been under water. Then if the Ozark wall were at the same time joined across the Mississippi, the four sides of the Great Basin would be completed. And when I survey all the evidence, it seems to me that this must have been the precise way in which the waters of this vast inland sea were confined.

Then for a more complete verification of this claim let us imagine a great mediterranean sea, more than 100 times as large as Lake Michigan, to have existed in this basin, and its waters to have accumulated on account of Eastern upheaval. We all can see that this vast stretch of territory is a veritable basin whose sides are of more than sufficient height, if filled with water, to form an inland sea more than one hundred times the size of Lake Michigan, more than 600 feet deep in the lowest part of the basin—i.e., in the region of the Great Lakes. If we could build a wall across the Mississippi, or rather restore the wall which countless ages have worn away, and again build up the mighty parapet that once stretched across the St. Lawrence a short distance below Quebec, a great sea would again accumulate. Step by step we would see the waters gathering in these two valleys. Year by year the broad expanse of prairie would become submerged, millions of acres of forests and numberless animals would become involved in universal death. Now, I hold that such an inland sea did accumulate over all this vast extent of land immediately after the New England mountains arose from the sea, and that this conclusion is supported by the most overwhelming evidence. Then, as before stated, let us imagine such a sea to have accumulated over a territory once teeming with abundant life, while we examine the evidence.

First, then, there are the three primitive walls on three sides of a great basin. Secondly, the fourth or eastern wall was reared across the only probable (may I not say possible?) drainage outlet. Thirdly, the greater part of this basin of more than 500,000 square miles in area, presents uncontested and incontestable evidence of having, in very recent geologic times, been the bed, over which, for unknown centuries rolled the waves of a fresh water sea.

A few facts may now be stated still further confirmatory of this view: Over all this territory lie entombed in a fresh water bed of recent origin, the remains of the mammoth, mastodon, and other huge pachyderms of interdiluvian times, while in the New England mountains there are none, save possibly here and there a single bone, carried perhaps by rivers from the basin into the ocean. This, it will be seen, argues that while these great quadrupeds luxuriated in the Great Basin Valley, the body of New England was sleeping in the sea. Again over this Great Basin Valley, are innumerable old river channels now filled with detritus, where no streams now flow, and which have been filled in recent times by over-towering waters. And again in almost every part of this basin, where examinations of these superficial deposits have been made, are found the remains of ancient forests, trees, stumps, limbs, leaves, seeds, grasses, etc., etc., plainly attesting that this covering was quietly deposited upon a vast area of growing vegetation. There is the buried soil; there is the vegetation it bore; and there are the animals that luxuriated thereon, all forever shut up in a mighty charnel house. Could this ever have happened; could these conditions ever have been brought about except in the manner here suggested? Thus. link after link added to the chain of evidence seems to banish every doubt, that there was, over this vast territory before named, long after the close of the last glacial epoch, a wide expanse of fresh water. After the glacial epoch, for the mud and silt was quietly settled upon a surface almost universally glaciated; and of fresh water, because of the total absence of marine shells, except as before stated in the lower part of the St. Lawrence Valley; and perhaps an occasional one carried from its original bed by transporting agencies. Again, where can there be found any other barrier to confine such a sea as all geologists admit gave rise to this superficial formation? We will search in vain for any other boundaries!

It seems then that the very presence of such a vast body of matter collected upon this area, must have, by actual mechanical pressure, depressed it somewhat, so that the surrounding ocean must have stood higher on the shores of the continent while that pressure existed than before or afterwards. Now this is a feature well known to geologists. Who is he that does not claim that the continent, or at least a great part, has recently been raised to a higher level? Can it make any difference whether the earth's interior be a molten mass, or a solid, plastic under the reign of implacable heat, when this transfer of mechanical energy from the continent and the ocean is accounted for? I cannot conceive how the measureless weight of a great mediterranean sea, could be removed from one part of the earth to another, without changing the water-line of the continent relieved of that weight. It is not far back in the geological history of the lower Mississippi, when the waters of the Gulf of Mexico reached the mouth of the Arkansas, and again retired, but to again approach as the ocean's waters were augmented. And one who is familiar with the features resulting from these great changes can, with but little difficulty, link them in order of time with the recession of these inland waters.

But we have now so nearly approached an unavoidable conclusion that but little is needed to reduce it to a demonstration. The great hypothetic sea has long since retired. Can we not find the tracks,—the way-marks of its retreat, and make them depose in support of our claim. Let us attempt it.

The great waters thus hypotheticated, I will call the Millerian Sea.* Considering the depressing effects it likely had upon its bed, it must, at the time of its existence, have received the waters from a large expanse of Canadian highlands. The Millerian Sea by some grand process made for itself two great outlets—i.e., the Mississippi and the St. Lawrence. Making due allowance for all likely depression, as shown by marine estuary deposits in the present river valleys, that sea must have towered from 700 to 800 feet above the

* I have thus named it in honor of my aged friend, Morris Miller, who many years ago directed my attention to the southern boundary of this sea, and who outlined it almost precisely as the late geological surveys have outlined the glaciated area. Now if this boundary be true for the glacier, it must also be true for the sea. Those desiring to learn further of the Millerian sea and the great floods attending the rupture of its boundaries, may obtain much from the author's lecture on the same in Volume II of the annular theory.

ocean. We can then faintly imagine with what terrific force its waters rushed to the boundaries of the continent when their ramparts by some process were ruptured.

We are now brought face to face with a question that apparently defies solution without the aid of "upper waters." How did this sea make for itself two outlets? Can we imagine a lake bursting its walls and rushing to the sea through two outlets, and continuing to cut down deep channels until it is drained? How did it ever occur that the St. Lawrence break accommodated the Mississippi rupture by not drawing the waters eastward and away from the latter? How did it happen that the Mississippi break did not close the St. Lawrence outflow, by drawing the waters thence? How did it happen that both breaks in opposite extremities of the boundary were made at the same time? Why did they mutually keep pace with each other, until the waters cut downwards and backwards channels for two of the greatest rivers of the earth? It is scarcely possible that in ordinary course of drainage the waters would not all have been drawn to one outlet. To account, then, for the two breaks and the two river systems, we are forced to admit that some vast and measureless supply of descending waters made the Millerian Sea to leap its barriers at both points, at one and the same time, and that that supplying-source kept up the waters so long that the excavations were far advanced under its government; after which each excavation continued independently.

But, if such a source supplied the retiring sea with waters, it must also have supplied a black sooty carbon, that settled to the bottom of the sea, forming the very superfice of the sea bed, when the waters receded. Now where must we find this carbonaceous covering, if it did fall? Certainly more abundantly in the northern, middle and western part of the basin. For the northern and northwestern slopes of the continent must have supplied it, for there alone were the rivers that could bear it seawards. And when it once reached the sea, the tendency would be for it to move with the moving waters toward the southern break. Then the carbonaceous matter, which I beg leave to denominate *carbonite,* must have settled more largely over the States adjoining the Mississippi.

Well, when one travels over the great prairies of the States referred to, he sees nothing more striking than the carbonite that covers this vast reach of territory. It covers all the hills, it fills all the swamps and sloughs; it is the foundation of all peat deposits, and it spreads over all the plains—a black top covering, varying from a few inches in thickness on the uplands to a few feet in the valleys.

I know I am now, as well as at many other times, rejecting popular opinion—that this black, superficial coating is the result of a slow accumulation of carbon from the annual fires, that probably swept over the region in former times; but while law presides in nature's high court of order, this cannot be so. While that universal and inexorable devourer, oxygen, is present in the atmosphere, every particle of unconsumed carbon arising from incomplete combustion, is afterwards consumed, nothing being left but the ash of vegetation. So that so far from being a carbonaceous product, black and pitchy as it is, the soil would rather consist of the mineral ash accumulation of centuries. We see this process continually going on around us. The dense, black column of unconsumed carbon rising from every locomotive, and chimney, is soon seized upon and dissipated. Besides it is found in the bottoms of ponds and lakelets, where fires did not devour, and where streams have not transported it from the surrounding regions. So surely then as a fire sweeps over a plain, leaving blackness in its path, so surely the unburnt carbon it leaves behind is re-burnt and made to disappear. But there are things that must forever set this question at rest. The carbonite when sealed from the atmosphere by a covering above it, is a purer carbon, and when dug up and exposed to the air will sometimes take fire spontaneously, but nevertheless leaves a black, ashy compound. This certainly proves that it had been covered and sealed from the action of the air ever since it fell, and never was the product of a burning vegetation. But if unyielding law is not sufficient to force compliance in one way, it may be in another. If the ten thousand lakelets and ponds of the great northwest on whose bottoms rests a stratum of carbonite, are not able to settle this question, there is one witness that none will fail to honor: Millions of boulders lie in and upon this pitchy soil. If prairie fires formed the black soil that covers the fields, they did not form that which underlies these lost travelers of a former day. Some boulders, when brought by ice floating upon the sea, were dropped upon a black, pitchy bed at the bottom of that sea. Thus, again, are we driven by the logic of facts to the eternal rock of Law, and the annular theory is settled still deeper upon its immutable foundation.

Here we find, also, lying immediately under the carbonite, the same kind of clay that accompanies the carbon deposits of the world. The same tellurio-cosmic dust of clay that accompanied every carbon downfall, and separating therefrom settled first because of its greater specific gravity. Now we may readily understand why, over so much of the great northwest, there is such a lack of forest growth. Is it not a fact within the comprehension of everyone, that if the treeless prairies were not covered by this seedless deposit from on high, they would be covered with forests as other lands? Is it not also a fact, well known and easy of demonstration, that whenever this seedless covering has been removed, there forests have sprung up? The rock-soil from which the oak, the hickory, ash, etc., invariably spring, has been covered by an impervious bed, seedless as the dust of space, and forest growth is an impossibility. There seems to be no other possible reason why the deep soil of the prairies is not as other strata. In short, it certainly is a fact, that if this deposit were the detritus of other and neighboring lands, they would be timbered as other lands. Here, then, is solved another perplexing problem.

A sufficient amount of evidence of sudden accessions of water throughout the vast lapse of time, during which the Millerian Sea was retiring, might be collated to fill a volume of itself; and it would be a pleasing task to give it now to the reader, but I must move on to other fields.

Look at the millions of valleys, channels and minor corrugations that have been made by the excavating power of running water! I can count fifty of them from my window to-day, through which no water runs, except during a rain. From yonder range of hills radiate deep channels that evidently could not have been made by such rains as fall at this age. Fifty years ago these hills were covered by the primeval forests, and rains could make no more impression upon them then than now. The autumn leaves gathered in these long trenches and hindered excavation. There the grass, shrubs and bushes are growing, and only when it rains, a powerless stream threads its way to the creek below. Did such transient puny streams make these deep gaps in the hillside? It cannot be. Yonder is a valley two miles wide, and the merest rill is the only excavating agent that occupies it. It is only one of thousands and millions that ramify in all directions the world over. And as I contemplate the puny agent and the grand result, I am forced to say that nothing less competent than appalling down-rushes of devouring floods could have made these streamless channels. And when I have stood before the grand old ocean, driving its devouring waves against the shore, and tossing its flowing mane on high, and have remembered that there are waters enough there to make one thousand floods, each of which would cover the entire earth fifteen or twenty feet deep; and remembering that these waters fell from "above the firmament" as fearful cataclysms; I see the world again and again writhing in the serpent folds of the deluge. I see man in the mysterious planning of Deity, the victim of immovable decree. Oh, Thou incomprehensible mighty One! Shall man's mortal eye ever penetrate this veil and read what lies beyond?

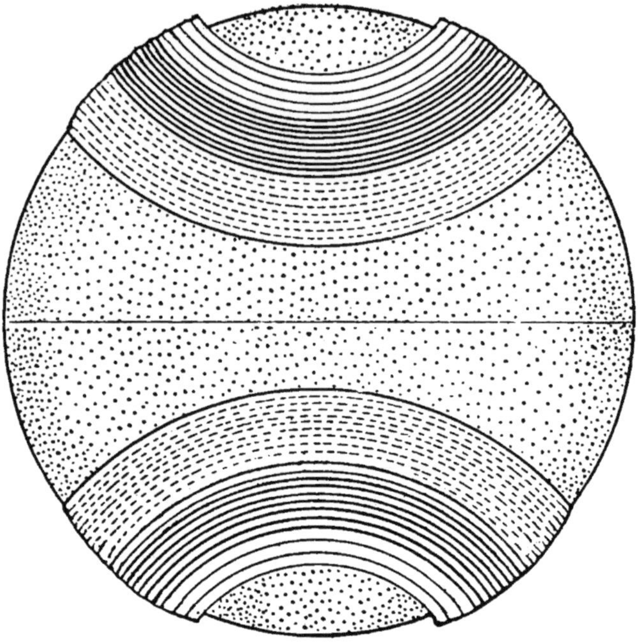

Fig. 6. THE CLOSING SCENE.
(EARTH WITH BELTS CAPPING THE POLES.)

Fig. 6 represents the earth stripped of its annular appendage and with its last lingering canopy suspended over the regions of both poles as vast clouds. Over the tropics and much of the temperate zones the vapors had become so thin that the clear sky could be seen at times and in places. The sun shone into this thin vapor sky and made it a most brilliant illuminator. The sun itself was dimly seen in this effulgent heaven as a conquering hero waging victorious contests with vapor foes. I have found this white and shining heaven with a hidden sun in the ancient thought of many peoples. This was the "Peplos" that Penelope wove in the day and unwove at night—a brilliant veil of vapors that illuminated the whole earth. But the God of nature had decreed that it should be taken down. and He destroyed "the face of the covering cast over all people and the veil that was spread over all nations." Is. 25: 7.

CHAPTER X.

THE GLACIAL EPOCHS AND EDEN RUINS.

ANNULAR SNOWS THE ONLY COMPETENT CAUSE.

Perhaps about 80,000 years ago,* the earth, now teeming with multifarious forms of life, was a scene of death and almost boundless desolation. The unmistakable language of the geologic record is that there had just closed a long era of perpetual spring.† The mammoth, mastodon, and a multitude of other huge quadrupeds, whose giant remains are found in the world's stupendous wreck, fed upon the products of a tropical and semi-tropical earth. Contemporary perhaps with these lived that race of beings upon whom we must look as the precursors of man. That was preeminently the age of huge pachyderms and other giant races. Their remains indicate that they were much larger than their living representatives of to-day. In looking over this pre-glacial—it may be inter-glacial world—the investigator is forcibly struck with its manifest completeness. It would seem that if there ever was an age when the earth came forth from the hand of the Great Architect in perfection, ready for the advent of man, and all that was necessary for his comfort and happiness, it was then. It was unmistakably a green-house world. The primitive elephant, and many of his congeners and contemporaries, fed in luxuriant forests, and grassy plains, where now the glaciers

---

* Geikie's "Great Ice Age," page 135.

† Belcher's "Last Arctic Voyage," Vol. I, page 380.

of the arctic world are holding them in relentless grasp, or grinding their bones to dust. How shall we account for this wondrous change—a comprehensive and universal change, so sudden and appalling as to leave upon the mind the impression that a far-reaching and all-involving destruction had overtaken the fair planet? This change is a well-known way-mark in the geologic past. Could the powers of heaven and earth, —the tornado and the earthquake,—combine in one grand revulsion to crush out the present life-forms of the earth, obliterate its cities, and cover in one vast rock-covering all that is now seen upon its surface, it could be but a repetition of the change that involved the pre-glacial world of universal life.*

Now, the geologist knows full well what the immediate cause of this great change was. He knows that, as the earth became peopled by an infinitude of living forms, under the influence of perpetual spring, in a tropical or semi-tropical world, so it became desolated by refrigeration, and the spread of snows and ice over the continents. These Titan plows,—glaciers and icebergs,—from the polar regions, again moved toward the equator, and continued to increase until almost every valley within the temperate zones was filled with ice. The glaciers plowed the plain, scarred the hilltops and carved the mountain side. Nay, hills were pushed aside by their resistless progress; valleys and river-systems obliterated and a living world made a panorama of universal death; in short, ground up and remodeled the surface of the luxuriant earth, for the introduction of new, but similar, forms of life.

It is a part of the labors of geologists to read and

*Geikie's "Great Ice Age," page 460; also pages 484 and 341.

study the "records," and give, if possible, a competent cause of these great revolutions. Many theories have been advanced in order to explain them, but few of them possess even the air of plausibility, and have been relegated to quiet oblivion. Among those having claims to our consideration, is that proposed by Dr. J. Croll, and which has the powerful endorsement of Geikie, in his admirable volume, the "Great Ice Age." It may be a lack in my power to comprehend it,—and yet there seems nothing puzzling in it,—but I am unable to see how a man of deep penetration can find natural law to defend it. To examine it in detail would swell this volume beyond its intended limits. I shall, therefore, state but few objections which I think must, in the mind of reasonable men, be fatal to it; and then advance the aqueous falls of the earth's annular system as the competent cause.

It will be necessary to explain some parts of the Crollian theory of glacial epochs to the common reader. It is well known that the earth's orbit is not circular, but in the form of an ellipse. So that in its annual circuit around the sun the earth once in the year approaches much nearer to that luminary than it would were its orbit an exact circle. Consequently, once in the year it recedes to a greater distance from it. The sun also being located not in the center of the earth's orbit, but, as it were, in one end of an ellipse, the earth whilst in the other end is far removed from solar warmth. Again the orbit is subject to exceedingly slow changes in shape, by which, in time, it is so far removed from the form of a circle that it becomes very eccentric, and the earth, of course, must recede to a vast distance from the sun. Now, Dr. Croll conceives that the globe, when in the aphelion part of its path, or farthest from the sun, accumulates more snows in its polar regions during its winters than the heat of summer is able to dissipate, which after ages of accumulation amounts to a glacial fund, and causes long periods of refrigeration or excessive cold. While this theory appears plausible at first sight, it is far from able to abide the test of analytical reasoning and philosophic law.

First: It ignores the law, long ago laid down by that prince of philosophers, John Tyndall, which may be briefly stated thus: Snows, to be formed, require the expenditure of solar energy, and the greater the amount of snows, the greater the energy required. To take the earth from the sun, then, robs it of snows, and of the possibility of the accumulation of snows. One would not think of increasing the working force of his engine by robbing it of fuel. I know there are a great many circumstances and qualifying conditions that may be pointed to; but under all conditions the fact remains, that, to cover the earth with ice and snow, you must increase rather than diminish the engine force.

Second: It makes almost an infinite number of glacial periods, in the vast ages of paleozoic and subsequent times, whereas they are few and definite, which both the silurian and devonian order of stratification abundantly declare.

Third: It makes the glacial periods regularly recurring visitations, while there is not the slightest evidence to be gleaned from the vast ages of geologic time that they did so recur. On the contrary, the evidence is that they came after long and very irregularly intervening periods.

Fourth: It is evident that a continent encased with ice by means of solar evaporation of the oceanic waters could never again become freed from its fetters; for, since it requires a great expenditure of solar heat to secure the formation of vapors, before snows can possibly accumulate, it is plain that the glaciers could not be melted unless the heat should become greater. But this increased heat would increase evaporation, and increased evaporation means, to some extent, at least, a greater precipitation of snows, and an increase of glaciers. The very energy required to melt the glaciers, is the same that would necessarily augment and perpetuate them. So that if a continent should once become refrigerated by increased vaporization how could it possibly become free from the grip of ice?

Thus in the very outset we meet most insuperable difficulties. We cannot expect the earth to become covered with snows by cooling it, and stopping the formation of aqueous vapor, and the sooner we abandon this most unreasonable claim, the earlier will the question be settled. Glacial theories have been rejected because they do not present a natural scheme of causation and sequence, and as it would be difficult to conceive of a theory more antagonistic to natural law than this one is, is it strange that such men as the illustrious Tyndall should hesitate to adopt it? Prof. Geikie says: "No half-explanation will suffice; the key which we obtain must open a way into every obscure hole and corner; each and every fact have full recognition in the theory which may be ultimately adopted." The consideration then of such difficulties as here presented, and which are far from obtaining even a "half-explanation," renders it strange that the Crollian theory should ever have received the support of such powerful minds.

If glaciers in all ages were always formed as local glaciers are to-day; if the vast continental ice plateaus that accumulated mountain high above the ocean's level in both hemispheres were formed in the same way as they are made to-day in the Andes, the Alps, and the Himalayas, then vaporization under solar energy went on synchronically with condensation and precipitation. But can it be possible that during the glaciation of a hemisphere, that hemisphere can be both warm enough to vaporize the aqueous element, and cold enough at the same time to build an ice-continent,—embracing millions of square miles? In order to produce the mighty ice continents of the glacial periods in the Northern Hemisphere, according to the current theory, one-half the earth must have maintained a genial climate, while the other had a temperature excessively arctic. We can imagine the Alpine glaciers to be constantly increasing by the vapors wafted over them from adjacent lands, warmed by solar heat,—the only way that glaciers now are formed,—but we cannot conceive of vapors carried from heated lands, by accommodating currents on a frozen world? To account, then, for the glaciation of the interior of continents, the snow and ice must have accumulated on its borders, and have flowed inwards and upwards from the oceans, which as all know was not the case. Hence it is conclusive that the glacial periods were not produced by glaciers formed as they now are formed. But there is no other competent cause for the accumulation of such snows than the decline of annular vapors.

Again, the well-known and peculiar properties of

glacier ice must always hinder its great accumulation, unless it accumulates more rapidly than it moves off. It flows, and it cannot be heaped up without limit. Its rate of motion is in proportion to the slope of its bed and the fund of ice. As water, by flowing, exhausts the supply, and cannot accumulate unless the supply is more rapid than the flow; so a glacier cannot increase unless the snows that form it are supplied more rapidly than it can retire. What, then, must have been the source of those snows that built a mighty continental ice-cap over the Northern Hemisphere during the last glacial epoch? With every opportunity to move down a thousand valleys and slopes to the south, or toward the seas, with every foot of increase in the depth of ice necessarily increasing its outward flow, I must claim that the earth has not now any source from which such a mass of ice could be supplied; and I am therefore driven to the grand and all-competent source of tellurio-cosmic snows in the earth's annular system.

As during the Noachian deluge the earth could have been desolated by surging and heaping floods from no other source than the "waters above the firmament," falling in medial latitudes; so we cannot expect to cover a continent with towering snows from any other source. Men of science must not conclude that glaciers always accumulated by the puny process that now builds a local ice-heap in a mountain valley. They must rise to a grander conception. The foundation of the glaciation of planets was laid in the igneous era. The implacable heat of the primitive earth necessitated the glacial epochs, and the present process of vaporization and congelation under solar influence is an insignificant process in the same direction by different means. Is it not a fact within the comprehension of all persons that if glaciers had no other source at any time than they now have, the arctic ice could never have moved over the Northern Hemisphere? Is it not a fact that they do not now accumulate in any land? The great Humboldt Mer-de-glace of Greenland, moves toward the sea, and the more rapidly the snow accumulates and hardens into glacier ice the more rapid is its motion coast-wise. So that neither in temperate latitudes nor in frigid climes can glaciers indefinitely accumulate by evaporation and congelation. But during the glacial epochs the tendency of indefinite glacier accumulation is apparent.* Therefore they did not accumulate as glaciers do now! This is the great enigma that puzzles so many.

It is the privilege of the annular theory to make this plain. Now it must be admitted that every drop of the terrestrial waters has fallen to the earth from tellurio-cosmic space! and more largely than otherwise these have fallen in polar regions! All that is needed for men to understand this is first to abandon the unreasonable and unnatural claim that these waters all fell to the earth in archæan and pre-glacial times; and admit the purely philosophic and natural fall of the same from over-canopying belts spreading and moving through the ages with a step as sure as the movement

* We have but to read such works as Agassiz's "Geological Sketches" to understand the immensity of the ice field that moved over the Northern Hemisphere during the great ice age. Glaciers accumulated in the St. Lawrence Valley several thousand feet thick. In their limitless sweep they towered over the New England Mountains, scoring and planing their rocky side six thousand feet above the ocean. I have seen their tracks indelibly chiseled 1,500 feet above glaciated valleys in the Blue Ridge. The same glacier that was urged up the St. Lawrence Valley no doubt filled the basin of the Great Lakes and the Mississippi Valley.

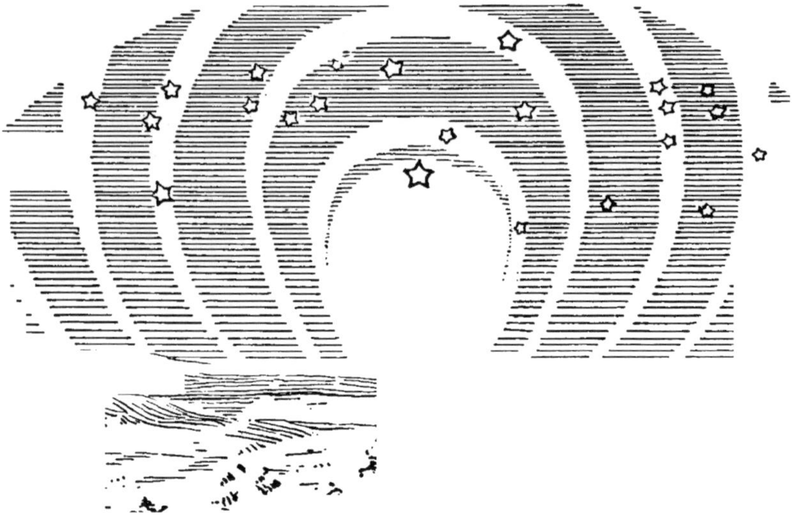

Fig. 7. ASTERIE, OR STARRY ISLE.

Τὰ δ' ἄστρα καὶ ἀρχὰς μὲν θολοειδῶς ἐνεχθῆναι.—*Diogenes Laertius.*

I have here illustrated what must be more fully explained in another volume,—the va enclosures of the northern sky, the last form of every falling world-belt. The Greek quot here taken from the writings of Diogenes Laertius is the "Moabite Stone" of th s problem, though translated variously, it simply tells us that the "*Arctic stars once revolved in a tho* Now a tholos is a vaulted enclosure—a space enclosed by an arched roof or dome. The au I have here quoted tells us that this was the doctrine of the earliest astronomers, and Anaxagoras as its advocate. But it matters not who first said that the "archaic stars revo (or dwelt) in a dome-shaped chamber" It affirms that the stars spoken of were north stars, for as surely as the earth had a canopy, as I have proved, man saw the stars first *dome-shaped enclosure*, and they were called "archaic stars" because they were seen among polar *arches*, not because they were *ancient* stars.

of worlds, toward the poles, where they descended as mighty and terrific down-rushes of snow. Men may call this the Vailian theory, or by whatever name they choose, but it must stand the test of law through all time, because:

First: It is natural, and it cannot be denied that the oceans now on the earth reached its surface as snow, at least to a great extent.

Second: It is evidently the only natural means by which great polar ice-caps could accumulate more rapidly than they could move or flow toward the seas, by means of which the earth became filled with snows and the continents ground and pulverized into mud.

Third: It is the only means by which great snow-fields could suddenly entomb a living world, which, as arctic lands demonstrate beyond a doubt, has taken place.

Fourth: It is the only competent means of explaining the presence of carbonite in great layers between strata of ice and snow, as seen in polar lands.

I have given sufficient reasons in foregoing chapters for advancing the claim that the oceans in great part fell at the poles as snow. I will certainly be allowed to make this claim at this stage of the investigation, considering the cumulative evidence pointing to the annular system in the whole line of discussion in the preceding pages. But let us now examine the *ipse dixit* of the polar worlds themselves.

Were we to turn our gaze upon the mighty wall of ice on the antarctic continent; washed in its grand circuit by the waves of the Southern Ocean, we would no longer doubt that such a mass of snows came upon that land from beyond the atmosphere. From what can be determined from discoveries by Ross and Wilkes, the imaginary "Antarctic Continent" is one mighty field of glacier ice, nearly 2,000 miles across. In many places where the glacier is washed by the ocean, it rises perpendicularly to enormous height, and extends below the surface of the sea to an unknown depth. It is an ice continent beyond the reach of snow-falls from vapor congelation. The vapors from warmer seas fall long before they reach this ice-field; and exist as impenetrable and almost perpetual fogs in latitude about 65° to 70° south.

Captain Foster, of the Chanticleer, spent several months at Deception Island, in latitude 65° south, and he particularly refers to this region of fogs. Though in the middle of summer, the air was so intensely cold and raw that some of his companions who had previously wintered in the arctic seas, declared they did not suffer more there than they did in the Southern Ocean. The fogs were so thick and frequent that for nearly two weeks neither the sun nor the stars could be seen. Here, on islands where almost continual fogs are encountered, glaciers might accumulate. These fogs are met with by all who sail over these waters, and are referred to by many voyagers on account of their prevalence. Here, it seems, in the circle of air between the line of eternal frost, and that of aqueous vaporization, the frozen vapors descend, as we would reasonably expect. I think it was Lieutenant Maury who reasoned from this fact, that aqueous vapors raised in warm parts of the earth never reached the polar world, but descended in the temperate zones. From what source, then, came this mighty casement of ice? Since it is unreasonable to suppose it came from warmer latitudes, since particles of redundant aqueous vapor must fall before the temperature of the atmosphere is reduced to that of the average in the polar worlds, it must be a mere accident that snow-storms ever occur in extreme polar latitudes. When sledge-tracks were seen by Dr. Kane nearly 20 degrees from the pole, though several years had elapsed since they had been made; when the bleaching skeletons of unfortunate explorers, articles of clothing, etc., are found uncovered in arctic snow, where it is *possible* snows might occur; when the bones of mammals such as the musk-ox remain for years exposed to view, as seen by arctic explorers, it seems indeed reasonable that snows seldom fall in the extreme polar worlds. How, then, did these boundless reaches of snow and ice accumulate?

Again, since it is well known and now generally admitted by geologists that glaciers did increase and accumulate, so that whole continents were covered by them; since there is no doubt that the arctic snows so gained upon glacier motion and decrease, as to push a stupendous field of ice through British America to its southern highlands, and then over these highlands into the basin of the Great Lakes and the St. Lawrence Valley, thus glaciating a great part of the North American Continent; and since such gains cannot take place except by and through sudden increase of precipitation, no more than a river could overflow its banks except by and through sudden precipitation; and since the great ice-flood was almost limitless in extent and in its effect, we are forced to admit that the source of the snows that formed it and forced it forward in its desolating march, was equally boundless, and the fall comparatively sudden and far-reaching. But such a source manifestly did not exist upon the earth. It seems to me as vivid as the noon-day sun, that if this earth had never been a molten mass, it would never have been carved by this mighty plow of the gods. For no such world of snows could have been formed.

And now, as we begin to examine the means by which the arctic glaciers were made the winding and burial sheet of an animated world, I would I could impress the reader with the majestic grandeur of the field before us. More than three-quarters of a century have passed away since by mere accident it was discovered that animals of a pre-historic period were entombed beneath and in the frozen soil and snows adjacent to the Arctic Ocean. Since the beginning of the present century, vast numbers of these animals, mostly entire, and remarkably well preserved, have been found, so that it is now a well-authenticated fact, that lands immediately under the arctic circle are to-day great charnel-houses of the interdiluvian dead. For a long time an extensive trade in ivory, dug from the frozen soil, was carried on by the Russian and Siberian traders, and it is still one of the staple objects of commerce in some parts of the frozen north.

An interesting account of some remarkable discoveries in this direction may be here introduced. The following is transcribed from the Penny Cyclopædia, Vol. IX., article Elephas:

"Mammoth bones and tusks occur throughout Russia and more particularly in Eastern Siberia, and in the arctic marshes. The tusks are very numerous, and in so high a state of preservation that they form an article of commerce, and are used in the same works, as what may be termed the living ivory of Asia and Africa. . . . Siberian fossil ivory forms the principal material on which the Russian ivory-turner works. The tusks most abound in the islands and shores of the frozen sea; and the best are found in the countries near the arctic circle; and in the most eastern regions, where the soil in the very short summer is thawed only at the surface, and some years not at all. . . . In 1799, a Tungusian, named Schumachoff, went to seek mammoth tusks near the mouth of the Lena. One day he saw among the blocks of ice a shapeless mass, but did not then discover what it was. In 1800 he perceived it was more disengaged, and in 1801 the entire side of the animal, and one of its tusks, were quite free from the ice. The summer of 1802 was cold, but in 1803 part of the ice between the earth and the mammoth, for such was the object, having melted away more rapidly than the rest, the enormous mass fell by its own weight on a bank of sand. In 1804 Schumachoff came to his mammoth, cut off his tusks and exchanged them with a merchant for goods of the value of 50 rubles (about $38.00).

"We shall now let Mr. Adams, from whose account the above is abridged, speak for himself: 'Seven years after the discovery of the mammoth I fortunately visited those distant and desert regions, and I congratulate myself on being able to prove a fact which appeared so improbable. I found the mammoth still in the same place, altogether mutilated. The Jukutski had cut off the flesh, with which they fed their dogs, during the scarcity. Wild beasts, such as white bears, wolves, wolverines and foxes, fed upon it, and the traces of their footsteps were seen around.

" 'The skeleton, almost entirely cleared of its flesh, remained whole except one fore-leg. The spine from the head to the *os coccygis,* one scapula, the basin and the other three extremities, were still held together by ligaments, and parts of the skin. The head was covered with a dry skin; and one of the ears well-preserved was furnished with a tuft of hairs. All these parts have necessarily been injured by 7,330 miles of transportation, yet the eyes have been preserved, and the pupil of the eye can still be distinguished. This mammoth was a male with a long mane. . . . The skin, of which I possess three-fourths, is of a dark gray color covered with reddish and black hairs. The carcass of which I collected the bones, is 9 feet, 4 inches high, and 16 feet, 4 inches long. The tusks (afterwards recovered) were 9 feet, 6 inches long, and weighed together 360 pounds. The head alone weighs 414 pounds. The skin was so heavy that ten persons found great difficulty in transporting it to the shore. I collected 36 pounds of hair trampled in the sand, by bears, etc.' This traveler goes on to state that the escarpment in which the animal was found, was more than 200 feet high, and made of pure, clean ice, and adds: 'Curiosity led me to ascend two other escarpments of the same material (ice), where I *found in the hollows great quantities of mammoth tusks, etc., of astonishing freshness.* How these things could become collected there is a question as curious as it is difficult to solve.' " (Italics mine.)

From my notes taken at the time the above was copied, I learned that this mammoth, while imbedded in the escarpment of ice, was 40 feet above the surface of the earth. Horses that have fallen in the crevasses of the Alpine glaciers, and remained hidden for many years , have finally made their appearance upon the surface of the ice. From this it appears that there is a tendency in the glacier to eject such remains.

A German traveler, named Erman, many years ago, visited the northern coast of Siberia, and has given some valuable information respecting that perpetually frozen land. In the summer-time the soil is thawed for a very few inches below the surface. I think it was he that refers to a well, sunk in the vicinity of Yakutsk more than a hundred feet through frozen mud and ice. In a great many places along the Siberian coast are huge ice-hills from 100 to 300 feet high, made up of alternating beds of ice, frozen mud, sand and carbonite, by some called peat. Imbedded in these masses are found vast quantities of mammalian remains, valuable for the immense amount of ivory they yield. The islands along the coast are apparently composed of these frozen fossils in a matrix of mud and ice. The island of New Siberia, more than a hundred miles long and thirty broad, seems to have been literally built of these materials, and whole cargoes of elephantine ivory are annually dug from their frozen hills.

Almost the whole territory of Alaska, so far as known, is covered with evident glacial deposits, and in many places mingled with abundant mammoth remains. In Kotzebue Sound, rise hills of considerable size, composed of ice of great thickness above the water, and extending below its surface to an unknown depth. On the top of this glacier ice, and covered with carbonite and mud surmounted with snow and ice, are found mammoth bones and tusks of remarkable freshness. In Europe, on the northern coast, these bones have been found in the same kind of deposits. Now as the tendeney of all these northern drift deposits is to move from the north, southward, it is evident that the arctic lands, near and about the poles, previous to the last glacial epoch, was as much the land of the living as it is now of the dead. But before I draw any definite conelusion as to the origin and cause of this great catastrophe we will still further examine into the nature and character of these remains.

Sir Charles Lyell says: "In 1772, Pallas obtained from Wiljuiskoi in latitude 64°, from the banks of the Wiljui, a tributary of the Lena, the carcass of a rhinoceros taken from the sand in which it must have remained congealed for ages, the soil of that region being always frozen to within a slight depth of the surface. This carcass, which was compared to a 'natural mummy,' emitted an odor like putrid flesh, part of the skin being still covered with short, crisp wool, and with black and grey hairs." Professor Brandt, of St. Petersburg, under whose care the above remarkable fossil has been made to speak to a modern world, says: "I have been so fortunate as to extract from the cavities in the molar teeth of the Wiljui rhinoceros, a small quantity of its half-chewed food, among which fragments of pine leaves, one-half the seed of a polygonaceous plant, and very minute portions of wood with porous cells, were still recognizable. . . . The blood vessels discovered in the head, appeared filled with coagulated blood, which in many places showed its red color."

Thus if we were cut off from every other source of information as to the character of this animal's habitat, this simple accidental circumstance of finding a part of a cotyledon of a plant in the hollow tooth affords evidence the most positive, and conclusive, that the race of extinct quadrupeds, represented by these frozen and mummied mammals, in the far north, not only was abruptly and suddenly overwhelmed by some mighty and immeasurable revolution of the forces of nature; but it also shows on what kind of food the animal fed, on the very day it was entombed, and that it was frozen up on the self-same day, and remained in that condition, as it were, in winter's eternal midnight, until found and dragged from its icy matrix. "In its stomach were found undigested fragments of coniferous wood," and therefore eaten a few hours, at most, before it was locked up in its wintry prison. Seeds undigested and so little changed as to tell plainly what kind of vegetation grew in the land where the animal lived and died, have frequently been found in the stomachs of these mammoths thus imbedded in ice and mud. "Even the capillary blood vessels," still retaining their contents, show that there was not the slightest decomposition in the body; all of which force upon us the conclusion that sudden and complete was the destruction that involved this wondrous race of pachyderms.

In the year 1843 Middendorf, of Russia, found several carcasses of these extinct animals, some of them remarkably well preserved. From one, the bulb of the eye was secured, and is now preserved in the Museum at Moscow. Among these remains were found marine fossils of northern species, characteristic of the northern drift, which shows that both were likely carried together from the north, in arctic glaciers. In the year 1866 many mammoths were found on the arctic coasts of Siberia, most of which still retained the skin and hair. They have also been found floating in icebergs out upon the open sea. One instance of this is related by Kotzebue, who was an indefatigable worker among the frozen seas.

By what natural means were these animals entombed? I trust this mysterious problem is approaching a philosophic solution. Lyell, from whom I have drawn largely in this chapter, looking back upon this great charnel-house of the mammoth, is forced to the conclusion that "the ice, or congealed mud, in which the bodies of such quadrupeds were enveloped, has never once been melted since the day they perished, so as to allow the free percolation of water through the matrix; for, had this been the case, the soft parts of the animals could not have remained undecomposed." This conclusion is arrived at by pure logic, and I cannot see how any man can avoid it. But what agency on earth, or in the heavens, known to man, could have thus involved them suddenly in ice and frozen mud, and kept them thus entombed unto this day? It is plain that a rise and rush of water over their forage ground is utterly inadequate to account for the facts. Consequently, a sudden tilting of the earth's polar axis, by which, as some scientists have supposed, the oceans of the Southern Hemisphere have been transferred to the Northern, also fails to explain the phenomena. The change of climate was sudden. In one day the animated races of the arctic zone, then supporting a luxuriant vegetation, were gathered down to the grave. Even if a transfer of oceanic waters could take place, this could not have changed the climate as by a stroke, and have congealed and sealed the land in ice; and no theory that does not agree with immediate and sudden change of climate,—that locked the mammoth in "pure, clear, glacier ice," without rush of waters and transportation of sand or mud,—can be accepted, as all debacles of urging floods leave the principal features unexplained.

Listen to the emphatic declaration of Cuvier: "If they had not been frozen up as soon as killed they must quickly have decomposed by putrefaction." Again let me call the attention of my readers to the remarks of this illustrious man, in contemplating the physical change the earth underwent,—by means of which perpetual winter involved the polar world. "But this eternal frost could not have taken possession of the region which these animals inhabited except by the same cause which destroyed them." A physical truth, more profound, and more in harmony with the annular theory could not be uttered. The same mighty down-rush of snows, from the earth's annular system, was the "eternal frost" that took "possession of the region which these animals inhabited," and, of course, was the "cause which destroyed them." Those tellurio-cosmic snows took possession of the whole polar world.

Having shown how necessary it is to admit that the cause was such as to produce immediate death and immediate refrigeration, let me ask the reader to find, if possible, any other competent cause than such a fall of snows. Now let us see how natural and how necessary such a snowfall is, and how admirably it accords with facts. The question to be determined is: What cause, suddenly and immediately destroyed the mammoth and his congeners, and froze them up in glacier ice, and kept them there till released by solar heat?

Is this proposition too exclusive? Since many of

these animals are really found in glacier ice "pure and clear"; since they have been found in icebergs in the open sea, and since an iceberg is a fragment of glacier ice; since those which are found imbedded in mud on the plains of Siberia, have evidently been dropped from icebergs, when that land was under the sea,—for marine arctic fossils are sometimes found with them (and if even one be found having these associations, it is strong evidence that all found in that region were once frozen in glacier ice, and floated thither in icebergs and dropped on the floor of the sea, among the living and dead fauna of the deep),—since the same may be said of the Alaskan remains; since it is well-known that these remains are fresher and newer near the polar sea, and must therefore have been released from the glacier in more recent times; and since, if they had been originally buried in the mud, without being transported, they must have been ground up by the great continental glacier; and that so many of them are entire, and well preserved even in the mud; and, finally, since their remains are yet being found in the glaciers on the Siberian coast, and frequently dredged up from the bottom of the ocean* whither they must have fallen from icebergs; we shall hardly escape from the conclusion that the mammalians now found fossil in frozen mud in the polar world or near it, were originally a part and parcel of the mighty moving glacier; and that as the last remnant of that glacier, the present ice-cap of the arctic world is still moving southward and downward and giving up its "mighty dead."

As we stand with this great problem before us, in the light of the annular theory every shadow disappears.

* Geikie's "Great Ice Age," page 300.

We see the mammoth frozen up in a glacier, and we know that that glacier was originally snow; and we also know he was frozen up in that mass, in the land in which he lived, immediately after he died; or, that he perished in his grave of snow, and has ever since remained there. Then it was *snow* that imbedded him in ice. That is, a fall of snow was the "cause that destroyed him"—i.e., was the "eternal frost that took possession of the region." Now, is there any possibility of escaping this conclusion? They are imbedded in ice that once was snow, and there they have remained from the day they fed on polygonaceous plants in that region. Then all that remains for us to do is to find a source of snow competent to supply it in sufficient quantities.

Since we know that every drop of the terrestrial waters must have come from the earth's annular system at some age of the world, and since there can be no doubt that some of those waters fell upon the earth in early historic times, and since it must be admitted that they fell largely as snow in the polar regions; and since, if not more than one-tenth of the waters now on the earth had fallen in the form of snow it would have covered the entire land-surface more than 30,000 feet deep; and since it can scarcely be possible that one-tenth of the oceanic waters did not fall in polar lands, as snow; the annular system comes boldly forth as a competent source of the snows that entombed the mammoth and his compeers. And when we see we are forced to find a source not only for a vast amount, but also one capable of affording it so suddenly and so rapidly as to suddenly change the climate and involve the land in death, we may emphatically declare that that source is the only one that can be found!

Imagine, then, those giant quadrupeds feeding in their natural habitat, until on a certain day, in autumn of the year, when seeds were ripe, the over-canopying belt of snows, having been approaching for ages that point where it is no longer moored to the skies, poised for a space in equilibrio, begins its downward course on the fated earth. In a moment's time a land that was rich and fat with life,—a world apparently launched in perfection from the hand of the Great Architect,—is chilled with eternal snow and frost. Inch by inch, foot by foot, and yard by yard, the snows fill the plains, fill the forests, fill the valleys, and chill the seas. Yesterday a world pregnant with exuberant life; to-day rocked with the mightiest revulsion, wrecked in the shock of ruin and disorder and discord, then wrapped in the white pall of universal death.

There may be some minor mysteries involved in the peculiar distribution of these animal remains in some parts of the drift, which at first sight will appear irreconcilable to the theory here advanced. These will vanish as the theory in its almost boundless conception is understood by scientists. I have no space in this volume to consider these in detail, and a brief consideration of them would be unsatisfactory to the general reader.

Having therefore taken the same position that intelligent geologists of all schools have adopted, and must stand to, viz., that the great catastrophe which overwhelmed the antediluvian animals at the poles and encased them in their icy tomb, was a *sudden* one; and that the same material that entombed them was the "cause or agent of their destruction," as Cuvier declares; if we follow this line of thought to its legitimate conclusion we are finally brought face to face with tellurio-cosmic falls of snow. For the material that entombed them is ice—glacier ice—and this was originally snow! And it was snow that fell upon the animated earth and froze it up. But such a fall of snows could not possibly have accumulated in the atmosphere as it now does and have fallen therefrom! Therefore it must have come from a source lying beyond the atmosphere. But beyond the atmosphere were the "waters above the firmament," recognized by primeval man. This recognition has been transmitted to us by tradition and mythology in such a way as to prove that a part,—the last remnant—of those waters, or rather vapors, fell upon the earth after man came upon the scene; fell too at that time as snows at the poles, and as a flood of water in warmer latitudes. But if water existed beyond the region of atmospheric clouds during that early historic period, or pre-historic period, if we choose to call it, then we know they were there in the mammalian age. That is, we know that while the mammoth fed on arctic vegetation there was a mighty over-arching fund of vapors, either frozen or otherwise, revolving around the earth,—a mass of vapors competent to involve the earth in ruin. What has become of those vapors? They are not there now; and the oceans are on the earth, and they stand deeper to-day, the world round, than they did before the last glacial epoch. If we deny the existence of this great abyss of snows and waters, so frequently referred to in the oldest documents and legends of man, we plant ourselves as opposed to law, and we will forever grope in darkness, and never find the true cause of the extinction of the prototypes of mammalian races, including pre-glacial, and, it may be, an inter-glacial race of men. On the other hand, as this investigation shows, if we admit the former existence of rings and belts of tellurio-cosmic matter which law requires, a thousand dark things are illuminated that otherwise meet with unsatisfactory explanation, or are wholly left in the night.

We may add strength to our position by other evidence. It is more than probable that if the mammoth was destroyed by a downfall of such snows as is here claimed, that previous to its fall, as in all former cases, —as in the case of the Noachian deluge, for instance,— the over-canopying fund of vapors acted as a mighty robe to the earth, keeping out the cold of space and confining terrestrial warmth, as well as equally distributing the solar heat over the globe. This equalization of temperature as in a mighty greenhouse, would melt away all existing glaciers and clothe the earth in verdure to the very poles. Let us not cast this conception aside without reason. It is plain that the solar beam of heat could reach the earth in a modified degree only, and earth would bloom as under a glass roof colored to prevent excessive light.

Now it would seem scarcely necessary for me to produce evidence to prove that the earth during the mammalian period did enjoy a greenhouse climate. The mere fact that these huge quadrupeds lived in arctic regions in vast numbers is proof that such masses of ice and snow as now are there were not there then. The mere fact that those animals were larger than their representatives of this age shows that they lived in a different atmospheric environment,—that the atmosphere was heavier, possessing more buoyant power by the mere pressure and presence of a vast ocean of vapors in its higher regions. But the ordinary reader may need something more than mere reference to these things. I will therefore briefly refer to some direct evidence.

Wherever we are able to find marine shells imbedded in the clays and other formations immediately below the glacial deposits they are known to be the representatives of warmer waters proving that warmer seas then occupied these northern lands. The presence of tropical and sub-tropical animal fossils in what are termed pre-glacial formations representatives of those now occupying the warm regions of Asia, Africa and America, shows that animals of the tropics migrated to the far north. Among the frozen cliffs of Siberia and Alaska are buried products of a vegetation of a warmer world. All these evidences exist in the superficial formations of Great Britain and Europe, as well as North America.* They are the index finger of time, pointing to a period of sub-tropical warmth in lands sleeping in eternal snows. But this is the very condition of climate needed to harmonize with the annular theory, and I take it as one more link in the great chain of evidence.

Now, as we have here laid before the reader a competent and, without doubt, the true cause or source of the snows that overwhelmed the mammoth and his cotemporaries, have we not also in the annular system a true cause and competent source of all the glacial epochs the world has ever known? This warm period

---

* See Geikie's "Great Ice Age," and any of the geological reports on the surveys of the States.

immediately preceding excessive refrigeration can be no accidental intervention. It was the product and result of eternal law. We find it, as previously intimated, again and again taking its place in its own proper order, in the harmonious gradation so conspicuous in the geologic column; and anyone must see that each repetition of it must add new strength to the edifice we are building.

But it is evident that other conditions must follow, which are so related, and dependent upon these stated previous conditions (as a warm climate followed by refrigeration), that we must attempt to hunt them up. I have shown that the natural associate of a downfall of snows from on high was a subsequent down-rush of water in medial, or extra-tropical latitudes. Now if we find in the record before us that immediately succeeding the great fall of the mammalian snows there followed a vast down-rush of devouring floods, the philosophic reader will not fail to see that the natural network of evidence is growing stronger and stronger, and wider and wider at each step. In the chapter on "Geological Topographical Evidence," the reader will find in addition to the evidence I now present the most overwhelming testimony naturally pointing to such correlative conditions.

I will first point the reader to the manifest evidence contained in what are known as the modified drifts along the valleys of the world. I say valleys of the world, for they are not confined to glaciated districts. Says Dana * : " The fact that such a flood, vast beyond conception, was the final event in the history of the glacier, is manifest in the peculiar stratification of the

* Dana's "Manual." pages 552 and 553.

flood-made deposits, and in the spread of the stratified drift southward along the Mississippi Valley to the Gulf, as first made known by Hilgard. Only under the rapid contribution of an immense amount of sand and gravel and of water from so unlimited a source could such deposits have accumulated." Again, "we learn that the region of the Great Lakes was probably one immense lake, and that the waters spread far south over the States. . . . The Mississippi waters, in the Champlain era, below the mouth of the Ohio, had an average breadth of fifty miles, and along by Tennessee and Northern Mississippi of seventy-five miles; so that it was indeed a great stream." This vast flood is supposed by Dana to have resulted from the rapid melting of the great expanse of glaciers; and he cites the spring floods of this age caused by the melting snows. But the melting of snows and glacier ice afford no similarity. Even spring floods are the productions of rains, as any one ought to know, upon fields of snow and ice, though the latter by its melting necessarily augmented the flood; yet it is known that the very presence of ice in the streams of a river system is a great safeguard to a more sudden rise of rushing floods. By it the flood is prolonged and modified in degree, and the presence of a continental glacier would not only check the natural or ordinary rainfall, but its melting, while it would vastly extend the duration of a flood, must have been slow indeed were it not for excessive rains. But imagine a rainy season over the surface of the ice fields of the present polar world! Imagine, if you can, majestic and sweeping floods occasioned by the melting of arctic glaciers. It is neither reasonable nor natural that the great ice sheet of the Northern Hemisphere should melt so rapidly as to fill all the tens of thousands of valleys radiating from it over the continents southward. And to suppose that such melting was greatly aided by ordinary rains is also unnatural, so that it seems evident that a flood in a glacial age requires a rain from a super-aerial source, and such a fall of waters must have been sudden and terrific. Besides, the very condition of the stratified superficial beds shows that the flood came and passed away before the glacier had nearly subsided. "There is direct evidence that the flood reached a maximum just before the close of the melting."* Now it must be plain that if this flood was occasioned solely by melting glaciers, unaided by descending waters, this fact could not be determined. The change in the deposits would be so gradual that no definite point of change could be readily determined. "The transition was a sudden one, as the abrupt transitions in the beds prove."* The same evidence may be gathered from other continents,—from almost every valley in the north temperate zone the testimony is the same.† So that it is not necessary to burden the reader with a detail of facts when he must see that the evidence is conclusive that before the close of the last great age of ice the Northern Hemisphere was swept by a stupendous deluge of waters strongly pointing to the source heretofore named. Those upper waters, even though they may have been in the form of vapor

* Dana's "Manual," page 554.

† I might largely quote from European authors in confirmation of this. I will refer the reader to the twenty-ninth chapter of Geikie's "Great Ice Age" for an exhaustive treatise upon this feature of post-glacial beds. also to Sir J. Lubbock's "Prehistoric Times," and fourth volume, page 305 et seq. of the Trans. Geol. Soc. Also Dana's "Manual," page 556.

in a greatly attenuated state, possessed the power of absorbing and diffusing the solar heat, and their very presence in the atmosphere in medial latitudes must have largely assisted in melting the ice, and perhaps for unknown centuries the falling waters and the melting glaciers deluged the land. But putting aside all other questions we see that the converging testimony of the continents is that an epoch of stupendous cataclysms of water followed immediately the epoch of down-rushing and all-involving snows,—the very thing the annular theory demands.

But there is one lacking feature that we must now refer to. It will be shown in another chapter how necessary it is that all such downfalls of water and snows when they have receded to the ocean and found their own natural level, must finally result in contineutal upheaval and mountain-making. Let us see whether this expected sequence follows in order. Suppose this triplicity of changes cannot in this case be established! Then it must be seen that the annular theory shall have met with a defeat. But if it be found without a doubt that the continents or even a part of them were lifted, attended or not attended with visible plication, I will certainly be permitted to take it as evidence corroborative of the views heretofore defined.

Let us first examine the coast of our own country, for bordering the oceans must the evidence be found, if found anywhere. In a great many places around the coast of North Amercia we find terraces of the Champlain period that must have been formed in the sea, for the reason that marine fossils are contained in them. At San Pedro they are now from 60 to 80 feet above the ocean, showing that the coast has arisen more than to that height since they were formed. They stand as conspicuous way-marks on the Pacific Coast. Also in the neighborhood of San Francisco there are many shell-bearing formations of this period more than 70 feet above the tide. Still further south, on the Mexican Coast, they rise almost to an equal height. There is but little doubt that an examination of the Pacific Coast will reveal many such beds between Oregon and Alaska. The eastern coast, however, north of Cape Cod, presents the most remarkable evidence of elevation in the quaternary at the close of the glacial period. It would require much space to specify all points. We will therefore epitomize from Dana,* who condenses from a vast field as follows: "The elevated sea-border formations that have been described *prove* that in the Champlain period the land where such formations occur *was at the water's level.*" (Italics mine.) From a critical examination of these beds by many competent geologists, from Lyell down, there can be no doubt that the New England mountains were elevated, immediately after the Champlain flood-beds were deposited, to the extent of 450 or 500 feet. The same authority says in another connection: "We hence learn that in the Champlain era salt waters spread over a large coast region of Maine, and up the St. Lawrence nearly to Lake Ontario." The sea was 500 feet deep at Montreal. The remains of whales and seals have been found in the St. Lawrence Valley, and a part of the skeleton of a whale was dug from the soil 60 feet above Lake Champlain. From these it is demonstrated that some portion of North America was elevated, and that mountain-making was a part of the work of the

---

"Dana's "Manual on the Quaternary."

Champlain period, which immediately followed the great flood of the glacial period.

Turning our attention to the Old World we find the same unmistakable evidence there. Around the British coast, and the shores of Norway and Sweden, there has been a general elevation of the land from 300 to 500 feet. Geikie, after reviewing the testimony afforded by an immense number of shell-bearing terraces in the various regions, summarizes as follows: "During the deposition of the clays and sands with arctic shells, the land, as we have seen, stood relatively at a lower level. . . . The sea continuing to retire, the British Islands became at last united to the continent."*

Thus it appears that men who stand highest in authority on these especial subjects without any particular theory to champion, have laid, as it were, the capping-stone demanded by the annular hypothesis. Its first demand was that there should be among the last of terrestrial revolutions a mighty down-rush of snows upon a tropical or sub-tropical world. It came, and it came, too, upon a world of life and bloom! But this demanded its associated downfall of water from the super-aerial "deep," and its demands were responded to through the open flood-gates of the skies. But these required that the foundations of the continents should respond to the measureless increase of energy exerted by additional oceanic pressure, through the evolution of heat and rock-expansion,—a response, a force which neither mountain height nor breadth of continents could withstand. Earth trembled at the demand, and the mountains lifted their heads to loftier heights. What theory heretofore advanced has explained these

*"Great Ice Age," pages 321 and 322.

things? When we see that the co-linking of these great changes necessarily requires an explanation not only of the wondrous changes themselves, but that it also demands a theory that accounts for the marvelous order in which they have transpired;—for the cause of this co-linking;—when we see the earth during a vast period of time characterized by a warm climate and a vigorous vegetation, just previous to refrigeration, precisely as in the Adamite period, and know that the latter was, with scarcely a doubt, caused by super-aerial vapors spreading from the equator to the poles; when we know that a spreading thus of vapors, while at the very time it must force a warm climate upon the earth, is but an inevitable preparation for a down-rush of snows, we can readily understand the reason of this order. And when we see the far-reaching and sudden result in the extermination of specific forms, in mountain making, in the advent of new forms; when we see this exhaustless energy used in upheaval, always directed against the continents from the ocean-world, we are forced to admit that some portions of the ocean's bed have been forced deeper into its plastic foundation by increased pressure,—the very thing the spreading and declining of annular matter must have effected.

Then, again, as we see this annular matter remaining in the firmament, and becoming the primary cause of the warm period of the Eden world, and the decline of that matter the cause of its destruction, it requires no strain of reason to claim that all such changes have had the same all-competent cause. Now the fact that such vapors did remain on high until the time of the historic deluge, and produced that deluge, warrants the conclusion that the post-glacial floods had the same annular origin. And as it is utterly unreasonable and unphilosophic to claim that the melting of great continental glaciers could form floods "vast beyond conception," unless succeeding canopies of annular matter forced a tropic temperature upon the frozen world, and as that matter was present in and beyond the firmament till the Noachian flood, the very means necessary to force such temperature existed at the very time that earth was wrapped in a mantle of snows and ice, and could not have descended from the annular form without over-canopying the earth, as with a greenhouse roof; I feel then that I am justified in the claim that the great telluric glaciers of recent geologic times were melted under the tropic influence of the annular vapors and accompanying deluges from that source.

It seems to me that the solar heat diffused among these upper vapors (after they once became a part of the attenuated atmosphere) by constant accumulation and radiation necessitated excessive rains in medial latitudes. The cold air moving from the frigid continents toward the warmer oceans and the warmer equatorial lands becoming laden with the moisture of those regions and returning in the circuit of currents must, it seems to me, have constantly deluged the lands adjacent the ice-fields; and no doubt much of the phenomena attributed to diluvial action in glaciated regions can thus be accounted for. However, one thing seems very evident: such deluges need and imperatively demand a tropical temperature, and a tropical temperature involving a frozen world emphatically demands an annular system!

Thus it appears that the very fact that the earth was even once glaciated compels us to admit, not only that an annular system did exist as its philosophic cause but also that such a system was necessary for the dissolution of the great ice fields. What feature of the "Great Ice Age" is there that does not confidently point us to that all-sufficient builder and destroyer? A ring of vapors and the telluric and cosmic dust it must have gathered throughout almost infinite time, at the beginning of the quaternary or mammoth period, declined from the ring-system over the equator, into the earth's atmosphere. As a pure result before explained, this ring as it declined into the resisting atmosphere, spread laterally into the form of a belt, and in its effort to reach the poles of the earth overarched the planet and formed a greenhouse or tropical world. Thus perpetually falling and perpetually widening, in the course of time its polar edges hung, as it were, betwixt heaven and earth, pass that point where they can no longer hang in air, and they at once, as if broken from their anchorage in the skies, descend to the surface. Upon the very fields of luxuriant and abounding life that the overarching vapors had necessitated, they fell as snow, and overran the earth. As unknown centuries pass away, amid flood and tempest they reach the ocean, and the latter climbs the receding shores, and the ocean's bed goes down but to force its plastic foundation under the continents. This intruding mass of molten or plastic mineral must lift the margin of the continents as no other force can, and the crumpling and rupture of strata in many places must give rise to volcanic phenomena. How strange, and yet how consistent and philosophic, that the exhaustless and measureless energy of the molten earth, transferred, I might say, to the annular system, and after millions of years had rolled away, be spent in making the aqueous crust of the earth,—lifting its rocky frame to mountain peaks by under-thrusts,—and sinking some portions of its crust to abyssal depths.

It is plain that if the fall of annular snows be the true cause of the last glacial period, as here shown, it is reasonable to claim that there must have been glacial periods of greater or less importance in almost all the geologic ages, and that the same cause essentially produced them all. We will, therefore, give a brief chapter upon this grand and exhaustless subject.

CHAPTER XI.

A BRIEF REVIEW OF THE GEOLOGIC AGES AND A PRESENTATION OF THE EVIDENCE THEY AFFORD OF PRIMITIVE GLACIATION, ETC.

Let us suppose that by some great change in the ocean's level the present continents, with their abundant evidences of glacial action should become covered by its waters. In the course of ages the way-marks of this age would become deeply buried under the detrital matter of other lands. The countless millions of boulders now seen upon the surface where the northern drift prevails, and denoting the track of the glacier or the iceberg; the vast beds of morainic matter; the assorted sands and clays; the striated surfaces, and the living vegetation, now so well known to the botanists, would become hidden under a covering of silt, which in the lapse of ages would become solid rock. Suppose, then, the eyes of the future geologist in an age to come should investigate this continent again heaved from the deep. What signs of our last glacial period would he see? Striations obliterated could afford him no evidence, and only where denudation and erosion should accidentally expose the buried drift could he find evidence that the quaternary was an age of ice. He might notice the paucity of fossils, an old drift bed, or peat swamp. With these he might find an occasional tropical plant, but with these he could come to no positive conclusion as to the climate. The plain evidence we now have of alternate warmth and refrigeration would be in great measure lost, or so confused that it would take a long time of careful and patient investigation, and comparing of testimony by the geologists of the world, before any definite conclusion could be attained. Such difficulties as these perpetually confuse the investigators of this day. In the grand carboniferous piles there is the most unimpeachable evidence that it was an age of tropical growth, in the lycopods, ferns and calamites of that era; but in the very heart of this evidence we find the track of the mighty glacier. In the permian age, amid its abundant evidence of a tropical life; in the Eden world of the tertiaries, the great ice king has again and again set his heavy feet. Then, just as we see in the last great ice age strong evidence of warmth at the very foot of the glacier we may expect to find this conflict of evidence in *all* ages. No wonder, then, as we look into the archæan formations we find the same difficulties. But as we admit the sudden downfalls of snow in the efforts of the upper deep to find its level—the ocean—we can readily see why we find so much evidence of death in the midst of life. As it seems impossible to explain these mysteries, so numerous in the quaternary, without the aid of these great cataclysms, so we will ever find it of the utmost importance to have their assistance in every geologic age. The great objection to these sudden changes has been the want of a competent cause; but a source having been found, not only a competent but a necessary one, there must needs be much less difficulty in the road of the geologist.

The geologist has never yet found the base of the aqueous stratified rocks. We know not how deeply these formations extend, and therefore know not how nor where to find the first evidences of glaciation. One thing, however, we do know; when the laurentian stratified beds were deposited there was an ocean on the earth. Then we also know that a part at least of tellurio-cosmic waters had fallen; and further that they fell according to law more largely in those parts of the earth distant from the equator. Still further, we know that it is possible they fell there as snows, and if not prevented by too great heat in the earth, these must have formed into glaciers, so that if we find among these laurentian beds even scanty evidences of ice action, we have reasonable grounds to predicate the existence of glaciers. For, as above stated, time obliterates, and we are justified in putting our magnifiers upon all evidence and drawing conclusions from the enlarged image. A single pebble known to be such, a boulder, even of small size, imbedded in rock, must be admitted as evidence. There are small areas of archæan strata exposed in various parts of the earth, besides the extensive ridge in British America. They are found in most of the great mountain ranges; as the Appalachian, the Cordilleras, as well as in New England, and in many places on the Eastern continent. In many of these places some evidence of ice action exists. In the laurentian of the Blue Ridge I have seen rounded stones more than a foot in diameter, firmly imbedded in the native rock, and strikingly different from the concretionary forms in the same matrix. In some places, as in the archæan of Michigan, as well as in the Ozark Mountains, there is an evident tendency to conglomerate formation. Among the laurentian quartzite are boulders of such dimensions as to require the power of floating ice to transport them. In some of the later archæan, pebbles became an important factor in rock making. These pebbles and boulders are derived from pre-existing beds, and many of the boulders a foot or more in diameter.* So that if conglomerate beds are evidence of ice action it is strongly evident in the later archæan. If the whole field were exposed to view, no doubt we would see some of the grandest monuments of urging floods and eroding glaciers; old moraines and polished pavements. But if we are disposed to doubt the legitimacy of such conclusions as this, from the evidence thus afforded, let us look a little further.

We have already had abundant evidence that aqueous additions to the oceans must increase the potential energies sleeping in mechanical pressure;—that every pound of additional pressure upon the ocean's bed adds additional heat to the underlying strata, already pressed to the point of fusion;—that every additional pound of pressure adds somewhat to the volume of melted matter, which additional rock expansion must force aside. Hence, though the ocean's bed may be hundreds of miles in thickness, the expansion of the lowest beds laterally must produce crushing and plication, and this must form elevation. And as we have before seen, we cannot conceive of continent-formation by any other process. When more than 30,000 feet of archæan beds were formed as sediment from the first ocean, it was that much of the mineral and metallic frame; each foot of thickness only adding weakness to the mighty casement. The lowest beds possessed a certain degree of heat as a result of mechanical pressure, and possessed certain dimensions resulting from that degree of heat. The waters at that time had their beds, and as those

* Dana's "Manual," page 159.

beds deepened, the mechanical pressure in such places increased on account of the gravitation of waters thither. This increase of pressure augmented the heat of the lower beds; this augmentation increased their dimensions and necessarily produced local plication. This occurring in the deep-seated beds naturally forced rocks between others, and this from necessity produced elevation upon the surface. Here the evolution of continents began. But this beginning was not until late in archæan times. This continent-making and strata-bending did not take place until the laurentian period closed; for the conglomerates and coarser beds which show violent agitation and movement of waters, lie unconformably upon the lower beds.* Now why is this order? Thus we see the laurentian proper closes at the very time the conglomerates are formed; and upheaval of continents occurred at the very time we would expect it, on the supposition that snows and deluges of water came upon the earth. Now we will see how this same order, beginning at the very time the grand structure lines of the continents were laid, continues unbroken throughout succeeding ages. A vast period of time rolls away, and many thousand feet of rock are formed over this first glacial deposit. A long quiet prevailed, and if no further downfall of water or other matter had taken place, there would, as I think, have been no more rock-folding and mountain-making, except in a local way. I mean if all the waters and their associated matter had descended to the earth in archæan or laurentian times, there could have been but one general upheaval and plication. We now pass from the quiet waters of the closing archæan across the boundary

* Dana's "Manual," page 159.

of paleozoic time. We enter almost immediately upon a cold and stormy age; for its first epoch is a time of boulder transportation and formation of coarse sand-beds. These are followed by finer beds, indicating succeeding quieter waters. The acadian epoch of Dawson, or the lower cambrian of English geologists, is still more plainly the time of glacial action than the archæan. Bordering on the archæan, in Clinton and St. Lawrence counties, also in Franklin Co., N. Y., the conglomerates occur in heavy masses. In some places beds more than 200 feet thick are found. In East Tennessee and North Carolina occur the Oconee conglomerate beds extending over a wide horizon. And according to Dr. Hayden, conglomerate beds lying on the archæan are found in the Black Hills of Dakota. In Northwest Scotland, in Lapland, Norway and Sweden, conglomerate beds are placed low down in the cambrian (lower silurian), where they lie unconformably upon the older rocks. Geikie, looking at these facts, is inclined to assign some of these conglomerates to a glacial origin.* And when we see in some places, as in the eastern part of North America, there was a general extermination of primordial forms of life, at the very time these beds were formed,† the evidence of refrigeration and glaciation becomes very apparent. For it is well-known that species that lived in the acadian or lower cambrian are not found in the upper beds, and in the former some crustaceans of enormous size became extinct forever; while no trilobites, so far as known, so common in the lower beds, are found in the next period. This appears to be the case not only

* "Great Ice Age," page 478.

† Dana, page 181.

in this country, but also in the lingula flags of the English division, showing that both hemispheres were likely subjected to similar changes from the same causes. There was, as the evidence seems to show, a far-reaching cause of extermination; and as this destruction of species took place at the time the great conglomerate beds of the lower silurian were formed, the evidence is certainly strong that paleozoic time began with the close of a glacial period. But we have seen that a glacial period necessitates a sudden accumulation of snows, followed by rushing floods; and as one glance at the conglomerate and boulder beds and their associates will convince the most incredulous that they show violent agitation and rapid deposition as well as catastrophic change, we are still further confirmed in the belief of annular downfalls. Compared with the succeeding beds of the Canadian and Trenton periods, no one can doubt a radical and general change of conditions; the latter periods indicating a long time of calm and quiet seas, and a warm climate over a great part of the earth; and the former pointing to conditions antagonistic to life, and evidently a step backward, preparatory for a grand leap in the line of progress.

But here again we are met with a decisive test. If these coarse formations were glacial deposits, or the result of urging floods from on high, then the oceans were augmented at the time they were formed, and upheaval and crumpling of strata must have immediately followed this increase of pressure upon the ocean's floor. It is with but little astonishment, then, but yet with a deep satisfaction, that an investigation of these hoary rock-volumes reveals the very test at the very time we need it. Geologists the world over know that such continent-making and crust-disturbance occurred then and there. The conglomerates and their related beds lie unconformably upon the lower beds, and in turn have similar relations to the overlying strata. The snows evidently came some time between the archæan and mid-silurian. The waters rolled away to the seas; the crumpling came, and the lingering glacier and the tottering iceberg, working as they do to-day, during unknown time, formed rock accumulations upon the upturned beds. Thus it seems that the archæan was closed by glaciation and flood, and that the resulting disturbances reached far into the cambrian, and that the paleozoic strata were planted upon the glacial ruins of a former world.

Now having closed a rough and stormy age resulting from aqueous and snowy downfalls, as the very nature of annular formation requires a period of quietude and rest, while the next ring is preparing during countless ages perhaps for its final fall, as it overcanopies the earth, it is plain that we must find the next phase of the silurian age to agree with this demand. We must expect a warm age, favorable to organic evolution; and as these are the very conditions, as announced above, in the Canadian and Trenton periods that followed this great disturbance, we can only add one more link to the great chain of evidence. Immediately following this great plication during the lower silurian there came a long period of comparative quiet. During this long age it is likely that icebergs from the polar world continued to transport foreign materials and drop them over the sea-bottom as at this day.

After many thousand feet of calcareous and silicio-calcareous matter were deposited, there came a time which we may call mid-silurian, when many species in the midst of life's full tide passed away. Vast numbers of lower silurian species never passed this gap in silurian time into the upper beds. This fact is well known. In peering into this remarkable hiatus, apparently in the very noon-tide of primordial life, well might the illustrious Dana exclaim: "This wide extermination shows change and catastrophe."* As all organisms of that age known to the geologists were oceanic forms, this catastrophic change proves a radical and abrupt change in the constitution of the terrestrial waters, which in the very start drives us to concede the fall of additional waters from on high. But this conclusion necessitates the augmentation of polar snows and a wide expanse of continental glaciers. If, therefore, we find the glacier's track and the dead forms of exterminated life side by side, the argument would seem to be complete, and we would look to annular downfalls as the cause of both.

Now the existence of boulders and conglomerated pebbles, the inevitable products of glaciation in the medio-silurian, is known to all. On the American continent the first step we take in upper silurian lands is among the Oneida conglomerated, a formation in many places literally filled with pebbles and water-worn boulders dragged from older beds. The Oneida conglomerate extends from Central New York, almost through the entire length of the Appalachian range. European geologists repeatedly refer to these boulders and conglomerates in the upper silurian. Dr. Dawson, of Canada, viewing the angular fragmental blocks in the Nova Scotia beds of this age, says: "These materials

* Dana's "Manual," page 212.

seem precisely similar to those which are at present produced by the disintegrating action of frost," etc. The heavy blocks and angular boulders in the Scottish highlands are claimed by some authors to belong to the inter-silurian age of ice. But this argument becomes most complete when we find that these features of inseparable extermination and ice action culminated in the tilting of strata and mountain upheaval, proving that an energy was added to the ocean's pressure. At that time the great Cincinnati anticlinal was formed. The Green Mountains were lifted on high, with their beds of the lower silurian upon them; and other parts of New England were then raised from the ocean. In Europe, also, were extensive upheavals, and coincident exterminations, so that I presume all geologists will admit that that age (medio-silurian) was one of extensive upheaval. Thus a remarkable triplicity of events is known, and each member of the union points to annular declension.

Again the glacier and the iceberg melt away, and a long period of comparative quiet, during which stupendous formations of calcareous rock were deposited, intervenes. The old silurian ocean grew warm, and new life-forms filled its waters.* Why did it grow warm? Why did many of its inhabitants perish as by a stroke, and new tribes take their places? Here is where the heavy dolomitic or magnesian limestone beds were formed. If their constituents were in the waters that

* It must be evident to the philosophic mind that new life-forms could not and would not take possession of our oceans, unless their waters were radically changed by a vast addition. Now while at the same time such additions would exterminate old forms as well as demand new ones, and said additions also cause upheaval and glaciation, it is plain that the annular system must come in as the great cause of all these changes.

fell in more primitive times, why were they not deposited long before this? If they were deposited long before, why did the waters take them up again and deposit them anew? And how did the waters work through the great mass of Canadian and Trenton limestones, refusing to appropriate *their* constituents in order to get at those more deeply buried beneath them? Now, as we are forced by other circumstances, it seems to me, to admit this oceanic augmentation, and are here again compelled to face some of the most unphilosophic and impossible propositions, if such a conclusion is not conceded, our claim that a new ocean had descended upon the earth, and from its waters the vast beds of dolomites were deposited, as so much of the great fund of tellurio-cosmic dust native in the annular system, is one that under any circumstances, it must be conceded, should be respected. When we admit another downfall in the medio-silurian, we can see the new environment that necessitates new life-forms, on the ashes of the old; we can see why the deposit was magnesian lime and not a carbonate; why there were signs of rushing floods and refrigeration; why there was curving of strata. Uncounted millenniums moved down the tide of time as this great limestone formation progressed; and as the waters became relieved of their burden other organisms came into existence. In some parts of the earth, the oceanic waters approached that condition adapted to the proto-typical forms of the vertebrate races; for even before we pass the boundaries of silurian time some of the ancestors of devonian life appear as timid pioneers upon the scene, while at the same time some coral and crinoidal forms become extinct. These changes—the sudden dying out of specific and even generic forms, and the sudden advent of new ones, without any intermediary link of relationship to existing forms—points to the annular system as the seed-bed of organisms. Of this more hereafter.

As geologists have long recognized the fact that there is a definite boundary line between the silurian and the next succeeding age, or devonian; and as I have before intimated that each geologic age was closed by a fall of annular matter, this boundary will be our field of observation. In many places the passage from the silurian into the devonian is quite abrupt. Throughout the Appalachian field the transition is marked by the presence of vast beds of mechanical sediment, indicative of great agitation of waters, and known generally in America as Oriskany sandstone. This formation is an extensive one, and in many places is so coarse as to be a veritable conglomerate, indicative of ice action. In the Old World there does not appear such positive evidence of ice movements; but in its place is other evidence of either a cold age, or a new supply of ocean waters. The deposits of the two continents are marked by a scarcity in many places of organic remains. In my own State where the Oriskany is widely lain, they are almost entirely absent. Now as the beds immediately below this formation indicate by their fossils warm and quiet waters, it becomes more positively indicative of climatic refrigeration. In the lower beds more than 300 species have been found and known, and what is more important and striking, they are in great part characteristically distinct from those above. That a wide reach of calcareous fossiliferous rock should be overlain by an equally extensive and sparsely fossiliferous sand and conglomerate formation it must be admitted shows catastrophic change,—flood and disaster.

Sometimes the shells of the lower beds are packed in great numbers in the upper, and this has caused some geologists to place the Oriskany in the silurian division. But its forming simply a passage-bed between this and the devonian, and laid down by ice and flood, will readily account for all these things. Thus as we move upward we see the same expected evidence of sudden and exterminating changes. If we consider the Ludlow beds of Europe as the representatives of this passage-bed—the Oriskany and Helderberg—and protected somewhat from the full effects of refrigeration, the gradation upwards is so far complete. And when we find the former in no small degree placed uncomformably upon the upper silurian, and the same unconformability on the Western Continent, we simply find the same adequate cause of crust-breaking and mountain-making at work. We find in both continents the full effects of stupendous additions of oceanic waters, and extensive continental uplifts immediately resulting. In many places, and notably upon the eastern border of this continent, the strata were upturned immediately after the first devonian beds were laid down. This places the operating cause of mountain-making in company with the extermination of species, the urging of floods and climatic changes, as in all other cases, and directs us to the annular waters.

There were many minor changes of this character in the long ages of the devonian, but we will pass over them all to the border-land of the so-called carboniferous, where we find abundant evidence of ice action in the passage beds of the two ages; not alone in the numerous beds of massive conglomerates, but in their wide spread over the continents. Dr. Dawson has said, in speaking of Nova Scotia: "In passing downward from the carboniferous, we constantly meet with unconformability." And Dana: "An epoch of disturbance of the eastern border region intervened between the devonian and carboniferous."* Great beds of conglomerate at the base of the carboniferous are formed into actual glacial moraines in different parts of Scotland.† Beds of sub-carboniferous moraines are authentically reported from Australia. I believe it is generally conceded that these are the work of ice, and geologists stand perplexed and amazed over the numerous morainic beds among the coal-bearing strata of America. Without a competent source of snows we might well feel perplexed in our efforts to explain the source of a glacier, reaching from the north polar world, before the birth of the Himalayas, to the Australian continent. If there be any reliance in the autobiography of transported materials, which make up these ancient moraines, that space of time between the formation of the upper devonian beds and those of the lower carboniferous must have been one in which the earth was severely under the dominion of the Ice King. Any one whose eyes are accustomed to fall daily upon the morainic matter of the great North American ice-sheet of modern geologic times, will not fail to put the consolidated boulder-drift found below and among the lower coal measures in the same class.

There are some facts connected with the carboniferous quite remarkable. One of these is the repeated oc-

* Dana's "Manual," page 289.

† Geikie's "Great Ice Age," page 479.

currence of conglomerated beds intercalated between seams of coal.* There are at least four coal seams opened in Pennsylvania around Shamokin, between five beds of conglomerate, in a vertical space of 250 feet; and in the body of the conglomerate itself at Pottsville important coal beds have been opened. But these are not rare occurrences. They occur in other parts of the United States. Also in Nova Scotia (Dawson), as well as on the continent of Europe (Godwin-Austen). So that it must be admitted that conglomcrates and coal are natural, not accidental associates. (Let the reader remember this very important fact.)

Now, reasonable men will concede the fact that these conglomerates necessitate the movement of glaciers and icebergs. But what a marvelous puzzle! Floating icebergs and continental glaciers in a hot-house world! glacial epochs repeated again and again in the very land where and at the very time a luxuriant vegetation was forming vast beds of coal!! and these all accompanied by rushing, impelling and devouring floods. "Geologists are staggered by these appearances," so says Geikie. And they must and will be sorely puzzled until they admit the philosophy of sudden downfalls of snow and other annular matter that brought each age to a close, and frequently characterized the age throughout to a greater or less extent. But be that as men choose to make it; here in the very field, bearing unimpeachable testimony to a warm climate, are multitudes of witnesses establishing beyond a doubt the fact of glaciation. We must reconcile these things by the administration of Law.

One glance at this complex system of carboniferous

* Report of Progress, Penna. Geol. Sur., page 628.

strata forces, it seems to me, two very important conclusions. The intimate, and as is often the case, the immediate contact of the remains of a luxuriant vegetation, with massive conglomerate beds,—the well-known products of excessive glaciation,—proves that the snows of that olden age came suddenly and repeatedly upon a world of life. From this again comes the necessary conclusion that the oceanic waters were greatly increased in volume, and we must therefore expect as a legitimate and necessary consequence a system of upheavals and strata-folding correspondingly stupendous and grand. Must I point my brother geologists to the well-known facts that support this latter conclusion? Need I tell them that the continents grown more stable with time would resist oceanic pressure longer, but when they began to move would move with grander strides? Need I point to the great convolutions of the earth's crust, known to have been formed immediately after the carboniferous beds were laid down? The geologists "know these things by heart." But to the general reader I must devote a little time.

It is manifest, even to a common observer, that the strata of the so-called carboniferous age were bent, crushed and folded into thousands of corrugations after the coal and conglomerate beds were formed, and before the permian wave rolled by. In the Appalachian coal field this is eminently the case. In this region the rocks to the inmost depths took part in the general plication. The stolid piles of laurentian, the universal casement of the world's silurian, confined under the heavy beds of the devonian, and the devonian beds themselves, all moved in response to the potential energy accumulated as the waters increased, and other matter from the annular system gathered upon the ocean's floor. It was emphatically a mountain-forming era. Some of the greatest mountain ranges were then lifted from their ocean beds. Speaking of this age, Dana, whose authority few will venture to question, says:* "There were no Alleghanies, for this region was a part of the coal-making plain; there were no Rocky Mountains, for these, as the carboniferous limestones prove, were mainly under the sea." Again he says:† "The coal period was a time of unceasing change;—eras of universal verdure, alternating with others of widespread waters, destructive of all vegetation, and other terrestrial life except that which covered regions beyond the coal measure limits."

Now when we reflect that such extensive exterminations, such great mountain upheavals, such great beds of conglomerates were produced while yet the fund of annular matter was unexhausted, where else can we look for the true cause of these changes? Why were such wide reaches of coal beds formed in the lap of the glaciated world? Well might geologists stand perplexed and amazed over such inconsistencies. In England and Scotland the same mountain forming went on, at or near the same time. We cannot close our eyes to this wondrous co-relation of phenomena. The light cast upon these questions in recent times is penetrating and dissolving the mists, and thoughtful men are now beginning to read the *ipse dixit* of the living rock; and will not cling much longer to the unnatural and unphilosophic theory of geological growth as now maintained. We have seen the grand and stupendous up-

* "Manual," page 355.
† "Manual," page 359.

heaval of continents, and the rearing of mountains that closed the so-called carboniferous age, and yet the carbonaceous deposits never ceased to form,—never ceased to fall as carbon dust,—till the last remnant of upper waters had descended to the earth after man was placed thereon.

Let us now take a glance at the permian, evidently a transition period of the carboniferous,—a time when the old life-forms began to decline and " run out," and a new environment demanded new ones, and in response to which a new page in geological history was written. But the permian, in order to be a time of such transition, must have been a time of great physical changes, whereby new environments were made; and conglomcrate beds found in strata belonging to this period plainly tell us that such changes did take place. We find in different lands these conglomerates made up largely of exotic stones, some of large size, and I have no doubt that if we could see the broad fields of the permian we would see a glacial period well defined. As it is, however, the rocks of this age are not so largely exposed as those of former ages, at least in North America. But in addition to conglomerate beds the permian has also large deposits of coal. Those who have examined stones of the former beds in fact, and who have expressed a sentiment as to their origin, I believe generally favor the view that they were carried to their resting ground by ice. Both Geikie and Ramsay are of this opinion. Geikie says: " Ramsay has given a detailed account of the occurrence in permian conglomerate of blunted and well-scratched stones, which seems conclusively to prove the existence of glaciers and icebergs." The same authority claims that the permian conglomerate of Germany shows a like origin.

These things being well-established facts we can no longer doubt that the permian witnessed one or more glacial periods. Then if our theory be true that such periods were brought about by measureless reaches and down-rushes of snow, the oceans must have again increased in volume and depth, and we must find evidences of this in the flexures and curves of strata, in connection with the glacial beds. As according to these indications in the carboniferous era and all previous ages the upheaval came, we will look as confidently for permian upheaval as we would expect thunder to follow the "lightning's fiery wing." "Murchison," says Dana, "remarks that the close of the carboniferous (including the permian, of course) *was especially marked by disturbances and upliftings*." (Italics mine.) Again, Dana remarks that all the country between the Atlantic and the Mississippi arose from the ocean in the permian period.* Again, "At the close of the permian there were great dislocations." And, again, the same authority says: "It is manifest that the period between the close of the carboniferous and the triassic was one of enormous disturbance." De Beaumont's "System of the Netherlands" includes the dislocations of the permian beds along the base of the Hartz Mountains; also those in Nassau and Saxony, which preceded the triassic beds, as well as many cotemporaneous disturbances in the permian of Wales and France. In the Ural Mountains permian flexures also occur; and it seems likely that the ruptures noticed in Australian permian occurred at the same time. These oscillations

* Dana's "Manual," page 368.

"show cotemporaneous movements on both sides of the Atlantic Ocean."*

Here let me call the reader's attention to these cotemporaneous displacements in both and perhaps all continents. How did it happen that over much of the American continent, and over lands separated from it by the expanse of oceans, elevations should occur, just at the time the snows and floods of the permian reached the oceans? Why a feature in such remarkable harmony with annular declension? But suppose those exterminations which our theory demands should here fail to support us. It would be a disastrous failure of the annular theory. Such witnesses will not forsake us. They crowd around us in greater profusion as we move down the great tide of geologic changes. The carboniferous cataclysms reveal the dead forms in abundance; they lie thickly strewn over the permian world, and new species spring, as it were, from their dust. Forms weakened and depauperated in the carboniferous age cease to exist before the close of the permian. Coral animals that formed oceanic reefs in olden time far more extensive than those in modern waters, so nearly perished that but a few straggling members of the great family Cyathophillus appeared afterwards; and some of the ganoidal fishes at the close of the permian had passed away. On the other hand, crocodilian forms came upon the scene, and finally almost possessed the earth.† The great family of *T*rilobites became extinct, as well as many mollusks. Many vegetable forms also died out. It was the last of the Sigillaria and Lepidodendra. Now it must be admitted that while

* Ibid., page 394.
† Ibid., page 371.

this extermination of old forms was in many instances a gradual process, we cannot avoid the conclusion that "at the close of long periods and epochs there were more general exterminations,"*—catastrophic changes that abruptly closed the life-period of many organisms. So far then as our investigations have been carried we find a most remarkable connection and co-relation of phenomena. We find there has been a regular routine of changes individually dependent upon the grand cause we have herein set forth. We have seen a fall of annular matter in the historic period, and we therefore know that the same kind of changes took place many times in the geologic ages, and when we examine these changes more minutely we see how harmoniously the indications point to the great conclusions here drawn. Not a link is missing in the great chain, and we must now see that our conclusions are in the main correct.

I might follow the geologic record exhaustively through these changes, from the close of the permian, through the triassic, jurassic, cretaceous, etc., and show that in each period there are unmistakable signs of recurring glacial epochs, by which wide and sweeping exterminations of species were effected. In each epoch is the evidence of boulders and conglomerates. Each epoch is brought to a close by crust disturbance, and each disturbance preceded by flood and violence; and we need no longer be puzzled by the fact that at each period the continents took a "plunge bath in the seas" before they were lifted to greater height. They were simply lifted higher by the waters as they sought their level. If this be not the true explanation of this

* Dana's "Manual," page 384.

puzzling enigma, where shall we look for another? To suppose that each time the continents are raised a few feet from the ocean they must first take this baptismal plunge is too ridiculous and unphilosophic for this enlightened age. And yet we read it as the deliberate conclusion of eminent men. The continents were indeed baptized and lifted from the deep, but under the eternal fixity—law. This digression will prepare us for an examination of the more recent periods. These things are to-day within the purview of all investigators. I have looked over the more obscure and difficult periods and ages, and shown their necessary dependence upon annular falls, and I will now leave the more apparent path for other men, and will draw my conclusion here by a rapid run over the remaining field. Some of the more recent periods were characterized by the most stupendous revulsions and upheaval. The Alps, the Andes and the Himalayas were lifted in these modern times. At the close of the jurassic period occured one of the grandest series of mountain-makings the world ever saw. Not only at that time were the Rocky Mountains lifted higher, but the great Nevada range, the Humboldt and Uintah Mountains, were heaved from the bosom of the Pacific, and all the continents were to some extent moved to their very centers. At the close of the cretaceous period they were again moved to their profoundest depths, and again we see the track of the glacier and the stroke of death. Beginning in the permian period the reptilian forms finally had unbounded sway upon the earth at the beginning of the cretaceous; but this period passed away with the almost complete destruction of mesozoic forms. Those monstrous members of the reptilian family, as the plesiosaurus and ichthyosaurus, fell victims of that universal catastrophe. The Old World was passing by with sudden and giant strides. Again and again the mantle of bloom and life overspreads the earth, and again and again it falls as the winding sheet of the giant dead. As reptilian forms gave way the ancestral forms of the mammal family (horse, elephant, ox, etc.) began their long and universal control of the earth. But through ice and flood they finally passed away. As I look over the ancient forage ground of the tertiary mammals, and see also the great telluric graveyards, where in immense numbers they have been gathered, doubtless by driving and sweeping floods, I am so vividly impressed with the suddenness and completeness of these visitations that it seems to me we need nothing more than a simple prospect of the battle-grounds and battle-scenes of the tertiaries to convince us that the claims I have here made are essentially true, even if no other evidence could be found. If some appalling flood should in this day sweep every animal from the Mississippi Valley, and bury it in the Gulf of Mexico, the latter could become no more really the home of the dead forms of this age than the great West and Northwest is to-day the burial ground of eocene, miocene and pliocene forms packed in countless numbers.

Now a short rehearsal may be in place. In the foregoing chapter I have labored to show that the glaciers of the quaternary were caused by a sudden down-rush of annular snows; and that from the condition in which the imbedded animals are found no other reasonable cause or source can be found. I then point out the philosophic and necessary consequence of such an avalanche from on high, such as specific extermination, wide flood deposits and rock displacement. We then find the same signs of glaciation in all the geologic ages, save, perhaps, the true archæan, and find also that such glacial evidence is accompanied by the same inevitable results of continental disturbance, mountain upthrusts, flood-baptisms and exterminations, and that at each of these visitations a new ocean is formed, as shown by new life forms. These natural associates we find continually in company, running through the entire series of ages, a four-fold cord of evidence that no argument can break—witnesses that no force can sever—while the general exterminations found attending them force the conviction upon us that our conclusions are correct. Now in the first place we cannot account for these universal exterminations of life-forms, whereby whole races of animals inhabiting different continents vanish, as it were, in the twinkling of an eye, except through catastrophic flood, or universal snows. In the second place, we cannot account for morainic and other drift diffusion except by rushing waters and moving ice; and, thirdly, we cannot satisfactorily account for the grand and universal uplifting of strata from the oceans as radiating centers of force, except through augmentation of volume and weight, by water and other exotic matter. We know that the forces that have upheaved the continents have always acted from the oceans, and that they have acted again and again, and we know that the oceans cannot of themselves give or use this force again, after it is once applied, except through force stored up as potential, either by additional waters or solid matter. Every boulder dropped by an iceberg in the Atlantic acts as an atom in the balance of continents, and so sure as the day comes when the ocean's weight in the scale becomes so great as to move the beam, its bottom must go down! It matters not whether the inmost depths are a molten sea; the pressure upon the sea-bottom can become so great as to fuse the deepest strata, and force it laterally under the continent. So that the latter are to-day, without doubt, supported by rocks or molten matter interpolated between the solid center and the surface by pressure from the seas. Thousands of feet of conglomerate and other foreign matter have been formed into beds on the sea bottom, and thousands of feet in depth have been added to the oceans. And when we see that many times the upward movement of the crust has been simultaneous in different continents, and occurred with a flood or ice period, we have, it seems to me, no possible way of avoiding the conclusions here drawn. Now when we take a comprehensive glance backward upon these numerous down-rushes of matter from on high we cannot wonder longer at the constant oscillation of sea and land throughout the geologic ages. We see forces employed competent, and causes adequate to accomplish these grand results, and we see them acting in harmony with law; in harmony with nature in every field, astronomical or geological.

Now as we look over the hotly contested field where the glacialists and anti-glacialists meet in interminable war, we can plainly see the real cause of the conflict. It is an honest difference. Since no competent source of snows can be found on earth, Prof. Dawson has declared substantially that a great continental glacier is a physical impossibility, and so long as such a source cannot be found, the professor stands upon a rock impregnable to all assault. But he stands a hero surrounded by fearful odds. A dozen glacialists, with Prof. Newberry, one of America's great geologists, at their head, have examined the great drift areas of the world with honest intentions, and their practiced eyes have seen too much not to know the glacier track, and the universal verdict is that it has been. Its indelible way-marks are carved as with an instrument of iron in and on the living rock. The blooming valley, the luxuriant forests, and the mountain's rock-ribbed sides, have felt its rude embrace.

Now this is the actual state of the question. A great continental glacier, without a competent source! I presume the great champion of the "iceberg theory" would be one of the very first men to admit the universality of glacial action, provided an efficient source of snows could be found. His powerful mind is too familiar with known facts not to see that this is now the desideratum, and while it will not bend to unphilosophic demands it will follow the dictum of law. The mistake of the glacialists is that they claim the earth now possesses an adequate source of snows when every feature of philosophic law is against such a conclusion.

But where are the glaciers that the super-aerial vapors must have formed as they gravitated to the earth? If both these parties could be led to see the irresistible conclusion demanded by the philosophy of the heavens and the earth that the oceans that now dash around the world could never have reached the surface of the planet, except by snow and flood, this otherwise interminable conflict would cease. *There* can be no terrestrial source of continental glaciers. The interior of a continental glacier could not be fed by snows from the oceans without the fraction of all law. The decline of the earth's annular system is, I think, in harmony with every requirement both of the glacial and anti-glacial schools. May I not challenge either one to present a feature it cannot explain? For nearly twenty years I have seen with profoundest regret these honest efforts of opposing parties, spent as they ever must be, so I think, in fruitless labor. It must be so. It must be true that every pound of the grinding, carving energy of the mighty glacier was stored up in the earth's annular system by Pluto's potential arm. The solar beam now only supplements that action in a continuation of the process ever circumscribed by law.

It is well known that the famous Sir Charles Lyell during the active part of his life urged upon geologists the important fact that almost from the dawn of recognizable changes in the earth, the change in climate, the extermination of species and the change of ocean's level were a triplicity of changes that always remained unbroken in the order of their occurrence. If there was a climatic change there was a new distribution of oceanic waters, and at the same time old organic forms died out, or were depauperated into races of underlings, and new forms came into existence under a new environment. This triplicity in the grand processes of world-making is most significant in the light of annular declension. Those great warm ages in medial and later geological times, when the world enjoyed a tropical or sub-tropical temperature, were followed by ages more or less arctic; and the world of animals that sprung into existence in the warm age gave way to species of a hardier type. Just as in the Adamic period races of weaklings gave way to more dominant types, so in all ages the same physical changes must have had the same far-sweeping causes. The warm climate of the adamite period, and the competent cause, before set forth are the grand keys that unlock this midnight mystery.

As in a greenhouse it would be impossible to keep a tropical temperature amid the cold of the wintry world without a protecting roof, so it seems to me it would be as impossible to convert this earth into a tropical world as it hurries on its way amid the more than arctic cold of interplanetary space, unless guarded by a protecting canopy of vapors, of which the annular deep was a necessary source. But the same protecting roof, under which the hothouse world brought forth its hothouse organisms, by its fall became the very source and cause of refrigeration that brought in hardier types, and mingled them together on the same floor.

When we see animals of a temperate or sub-temperate climate solidly frozen up in eternal ice, how can we but conclude that here is the adequate cause of the climatic changes that have so frequently swept over the earth in the geologic past? The record is plain. The organisms of a warm climate were crushed and buried under the heel of winter. Winter could not have set his heavy foot upon a tropical land except through a fall of snows in the polar region. Neither could the mighty grip of the Ice-King have been softened, and the universal sweep of glaciers transformed into urging floods, except through the involving canopy. And now when we see the unmistakable track of floods, "vast beyond conception," at the very time the glacier stretched its naked front across the continents, need we longer marvel at this triple order of changes? A vast debacle of snow and water, continuing for unknown centuries! The pure result, a new distribution and condition of oceanic waters, refrigeration and consequent extermination of species!

If the illustrious Lyell had but caught one glimpse of the cause of this co-linking of changes so comprehensive, what an imperishable monument that master-mind could have reared for the admiration of the world! See the limitless snow-fields, covering much of the Northern Hemisphere, filling the valleys, and towering over mountains, until one vast winding-sheet hides the living continents! The revolving vapors having fallen, the sun shines down upon the snows, but as powerless as it now shines upon the glaciers of the Alps or the Himalayas; the ice-cap could not by this have been softened into floods. But, lo! Another telluric ring gradually descends into the atmosphere and over-arches the earth, and the ice-bound continents come directly under the influence of a hothouse temperature. Glaciers could not but soften and melt into deluges, "vast beyond conception," in such a world. Time rolls on; the great glaciers are transferred to the sea. A measureless pressure is lifted from the continents, and placed upon the ocean's bed, and that bed under the beck of law expands! The plastic matter against which that bed is planted is forced latterly under the edges of the continents, and these rise in the balance of energies. If this be not the process by which this order of changes is brought about, where else shall we look for efficient causes? Where else shall we find a philosophic reason for the lifting of continents, the folding of strata, and outbursts of liquid rocks at the very time of climatic changes? At the very time old forms die out?

At the very time the ocean's wave leaves its accustomed shores and rolls to other lands?

The well-known fact, then, that majestic and far-sweeping floods closed the last glacial epoch, it seems to me, forever demands a rapidity of ice dissolution that the solar beam could not produce through the clear atmosphere of to-day. And since we must, at the same time we are looking for a sufficient cause for this rapid melting and overwhelming water, also look for the grand cause that made a tropic climate succeed refrigeration, the grand cause of exterminations and mountain-making, the task is infinitely more difficult to explain without the aid of terrestrial ring-falls. If geologists then would simply admit this little fact that the oceanic waters could not, and did not, all descend to the earth in primitive times, the mystery will vanish at once. The great panorama of terrestrial changes will unfold; for it must be seen that such waters upon reaching the earth in after ages could not but cause refrigeration; could not but cause excessive floods; could not but cause extermination of specific forms; could not but cause new distribution and conditions of oceanic waters, and finally could not but cause crust-folding and crumpling of strata.

CHAPTER XII.

EVIDENCE ADVANCED IN SUPPORT OF THE CLAIM THAT THE EARTH'S ANNULAR SYSTEM WAS THE SEED-BED OF ORGANISMS, AND CONSEQUENTLY A REGION OF MICROSCOPIC LIFE AND INFUSORIAL FORMS.

Not many years ago I clipped the following from a newspaper after it had gone the rounds of publication in some of the leading magazines, and was accorded by them the importance of an established fact:

"The somewhat rare phenomenon of a fall of golden-yellow snow occurred in the midst of a severe storm on the afternoon of the 27th of February, at Peckeloh, in Germany. A specimen of the water melted from the snow, after being kept a few days, was microscopically examined by Weber, who describes it in the 'Wochenschrift.' He found that it contained principally four diferent kinds of germs or organisms, shaped respectively like arrows, coffee beans, horns and dark flat discs."

In the year 1846, Tenth month (October) 17th, fell the memorable shower of microscopic organisms, near Lyons, France. This shower of microscopic germs contained more than one hundred different organized forms, so different from anything terrestrial, as even in that day to suggest to many of the French savants a "cosmic origin."

A shower of strange organisms, described by Darwin, which fell near the Cape Verdes, covered more than a million of square miles, thus proving by its vast extent

Fig. 8. SUN WITH PERIHELIA.

Here is an attempt to approximate the general features of the solar forms that must have whirled their daily course across the vapor heavens,—"A flaming sword that turned every way." These cherub features necessarily accompanied the knowledge-giving tree, and at the same time were most effective world-guards of the way of the life-giving tree. While they flashed amid the vanishing vapors, neither man nor any living thing could partake of or behold those life-prolonging features that thus became a happy type of the Life and Way.

It is now freely admitted that the sword and the cherubim of Eden were solar features of some kind. When, then, we admit the canopy as the true solar vicegerent, we see how these perihelia must be allowed.

that it had an origin beyond the limits of the terrestrial atmosphere.

In the year 1803 a great shower of exotic organisms fell over a vast territory of Italy and Southern Europe.

In the year 1813 a shower of organic germs fell in Calabria, and from specimens subjected to microscopic examination sixty-four different species were obtained.

In the year 1755 one fell in Northern Italy, "covering about two hundred square leagues," and covering the earth in places to the depth of an inch, and at the same time a shower which reached into Austria fell as colored snow in the Alps to the depth of nine feet. As modern researches have proven that colored snows are filled with organisms, giving the color thereto, we can scarcely imagine the immensity of organic matter in such showers, and can hardly conceive it to have had a terrestrial origin. Homer speaks in the Iliad of one of these organic showers. In Northern Europe such snows have frequently been seen, and sometimes they have been accompanied with carbon dust, which must have had its origin outside of the realm of atmospheric oxygen. A few years since such a shower fell in Western Kentucky. They also have been seen in Asia Minor and Palestine. Microscopic examination of polar snows shows that they are permeated with organisms and particles of cosmic dust. Dana has said the origin of these organisms and dust is unknown. One significant feature in the case is the great similarity of the organisms in all showers; more than 300 different forms have been determined by microscopists, and all attempts to find a terrestrial origin for them have failed. If they had been taken up from the earth in whirlwinds and tornadoes, the fact could easily be proven. Those which fall in Europe, one would suppose, would be traced to Africa, but the species so far examined prove that they do not come thence. Out of the 300 species only fifteen have been found in South America (Dana) In the showers above mentioned (1803 and 1813), the two had twenty-eight species in common, or about one-half. Says Dana, the "zone in which these showers occur covers Southern Europe, Northern Africa, with the adjoining portion of the Atlantic and the corresponding latitudes in Western and Middle Asia."* Some decisive tests are well-known—a common origin and a common region of declension. A common origin, because of so many species in common, and the impossibility of finding a terrestrial source show that these organisms have their source either in the atmosphere of the earth or beyond it. If we admit that source to be in the terrestrial atmosphere, we at once contravene eternal law; for in that case we presuppose that the vast amount of matter descending as dust showers, filled with organisms, was made in the atmosphere, which could not be. The shower that fell at Lyons, in 1846, was estimated by Ehrenberg at 360 tons, and 45 tons of it consisted of organic matter of more than 100 different species. Well might Dana exclaim, after considering the numberless instances of such falls: "With these facts before us, how many millions of hundredweight of microscopic organisms have reached Europe since the days of Homer?"

Now I suppose some of my readers will hesitate to admit that organic germs exist co-extensively with mat-

---

* Dana's "Manual." page 634; see also Ehrenberg's "Blood Rains," 1847.

ter. When we know that there is not a spot on the earth, nor within the earth beyond the limits of present or former igneous activity that may not possess organic germs; scarcely a spot where life of some kind does not exist or has not existed; when we see the atmosphere everywhere contains floating germs; that in hot springs and in polar snows, in the lava-built walls of volcanoes, and even in the massive beds of the archæan rocks are found the remains of life-forms, it seems to me we are forced to admit the universality of material organisms; that they are co-existent and co-extensive with matter. Can we even imagine a nebula, anywhere floating in the immensity of space, that does not contain life-germs in that potential condition necessary in the evolution of life? Because we are familiar with certain life-forms, life-conditions and life-habits in a completed world, we are not to conclude that other forms cannot exist in an incomplete one. The dust showers that from time to time fall upon the earth, then, have either a cosmic or tellurio-cosmic origin. They may be, it seems to me, micro-cosmic clouds moving in interplanetary space, which meeting with the earth in its path are precipitated upon its surface. Since the history of the earth reveals the fact that it has been in many conditions, and each condition has had its peculiar forms of life organisms; and since every environment now has its own forms adapted thereto; and since, as each environment becomes modified or changed, the organic forms are also changed to harmonize therewith; we see it to be the declaration of law that almost every condition of the earth has had its peculiar forms of organic matter. Therefore in looking back to any condition of terrestrial elements, save perhaps that of plutonian activity, we must predicate some forms of life. Is it more probable that the omniscient Dispenser and Planner would place life-germs, or living forms, in an ocean of water on the earth than that He would plant them in potential attitudes in those oceans before they reached the earth's surface? Take into our minds, for instance, the last or outmost vapors that fell at the time of Noah. Having remained for unknown millions of years on high, receiving constant addition of meteoric and cosmic dust from without, and having originally received material distillations as vapors arising from the primitive earth, can we upon mature reflection conclude philosophically that that revolving fund of matter was not filled with organisms as surely as the gaseous envelope that forms our atmosphere contains them now? I, for one, cannot conceive of matter in a nebulous condition, which is not yet pregnant with life; since life is the original and primal force-element that developed into all we now behold. Can any philosophic mind, familiar with the grand history of this planet's evolution and development of its life-forms, under the intelligent direction of a God of law, look back upon the earth's annular system then, and not conclude that it was a region of microscopic life and infusorial forms? Can we look out upon the annular system of Saturn with a Designer its eternal Pilot, and disconnect it with primordial life? Can Jupiter's belts, necessarily composed of the elements common in the frame-work of worlds, illuminated and warmed by the electrifying and vitalizing power of the solar beam,—heaven's material vicegerent in the development and maintenance of physical life,—roll through space for millions of years, lifeless and spiritless?

Now if Jupiter's belted system of mineral, metallic and aqueous matter should have long ago descended upon the planet's surface, and we could see his continents and oceans, as we now can see those of the planet Mars, we would conclude that animals trod its continents, swam in its oceans and flew in its air,—in a world of completeness, a congeries of completed life-forms. Then we are only following a line of philosophic reasoning to conclude that a world of incompleteness must contain incomplete or primordial life-forms—forms that must in time develop according to the design of an Omnipotent Planner in the beginning. We see, then, in the organic yellow snow-cliffs of Bathurst and Greenland, in the dust-showers and "blood-rains" that reach the earth, evidence that finally leads us to the conclusion that organic forms are the natural accompaniments of the nebulous and elementary forms of matter. As we move along in this discussion we will see how true this must be. If such were not the case, then the eozoön canadense and its related forms were made as an immediate and separate creation after the archæan ocean fell to the earth, and following this the silurian forms in all their variety formed an individual and separate creation after the silurian annular matter fell. If these life-germs were planted in the earth alone, after the oceans declined to its surface, then they were planted there in all their potentiality at that time; and the whole scene of organic evolution would be one uninterrupted transition, without a "break," without a "gap." The rocks would contain all the transitional forms, or so-called "connecting links." But how very different from this are the facts scientists are well aware. The "breaks" are there; the "gaps" are there, and the geologist cannot fill them. The "links" are not there, and he can never supply them. Here now we have room and material for a volume on organic evolution placed upon the only philosophic foundation that was ever laid for it—an evolution from mona to man under the guide of law that began in the earth's annular system and terminated in its fall. As I cannot follow these ideas very far without swelling this volume beyond proper limits, I must refer the reader to Volume II of this series.

Some of my readers have doubtless heard of the "atmospheric spider." These little insects have legs, and apparently all the organs that the common spider has, but its home is in the air. It has the power of throwing an exceedingly subtle spider line with an actual float on the end of it that rises like a balloon in the air, and the little animal clings to its web or line, and floats off in the atmosphere. I was once authentically informed by an eye-witness of the fact that during a certain condition of the atmosphere countless thousands of these spiders descended to the earth, like a real shower of cosmic dust, and that for several hours they existed in immense numbers on the earth. However, toward the evening of the day, these all had disappeared, and they were noticed to spin lines which floated on the air as above, and then clinging to them took their flight into the air. I have at different times beheld real gossamer showers;—countless millions of lines descending from the heights of the air, so thick in the atmosphere as to counterfeit the glare of the sun.

My readers will likely remember reading in White's "History of Selborne" his account of gossamer showers which he had seen. One in particular which continued to settle for nearly a whole day on the earth, all coming down from unknown heights.

Darwin saw a spider shower in 1832 off the coast of South America, 60 miles from land. He also spoke of their spinning a floating line or lines, which, balloon-like, bore the little aeronauts aloft. A writer in Chambers' Journal says: "These gossamer showers are great mysteries, the air on these occasions becomes literally crowded with tiny parachutes composed of a few threads of almost invisible gossamer, each of the parachutes being occupied by a Lilliputian aeronaut in the shape of a very small but active spider." This same writer speaks of having seen one spider shower in 1875 and another in 1880. He says: "Fixing my eyes upon one of the spiders, I observed that as it left the gossamer pathway it selected a clean spot on the iron railing and gathering its limbs closely together, it projected from its spinnerets several threads which expanded outwards and stretched upwards from nine to twelve inches. Then this parachute seemed to show a buoyant tendency, and suddenly the tiny creature left hold of the iron rail, or was lifted off it, and quickly 'vanished into thin air.'"

Dr. Martin Lister, on one occasion, in York City, went to the top of the Minster, and from that lofty height still saw spiders there descending from heights unknown.* Here is evidence sufficient to prove that in the heights of the atmosphere beyond the points perhaps yet reached by man lives a race of spiders which are as much at home in thin air as man is on earth. There they live and propagate; there exists their food. Thin air is their peculiar habitat, as water is the home

* "Friend," Vol. LVI, pages 172 and 173.

of fishes. Now if in such a location can live little animals in countless numbers with legs and perhaps all the organisms peculiar to spiders, we may reasonably conclude that a cloud of vapor and cosmic dust floating in interplanetary space is also the abode of some form of life. If a spider can live in thin air, and also descend and live a while on the earth's surface, it could live in a nebular or a planetary belt. If a toad can live for unknown ages immured in solid rock, where neither air nor food can reach it, it could live, I presume, in a revolving belt of aqueous and mineral matter.

Now I am not going to advance the claim that batrachians or spiders lived in the earth's annular system. But I must claim, however, that the manner in which specific living organisms have succeeded each other on the earth as revealed by the geologic record demands that that system was the cradle of infant life,—the propagating beds in which the life-germs were placed by the great Gardener of nature. Men may laugh at this, but it is not half as ridiculous as to claim that all life came from monera or the rizopods in the primeval ocean on the earth. It is just as reasonable to suppose that germs took form in the waters, under the Creative Hand, before they fell to the earth, as afterwards, and when we see that each and every downfall brought in new life-forms which exhibit no specific relation to previous forms, we are forced to admit that either the seed-beds of the annular system provided the undeveloped organisms or there was a special creation at each period. We are, I say, obliged to admit this, or concede the hiatus separating age from age, and form from form. We are obliged to admit this, or forever con-

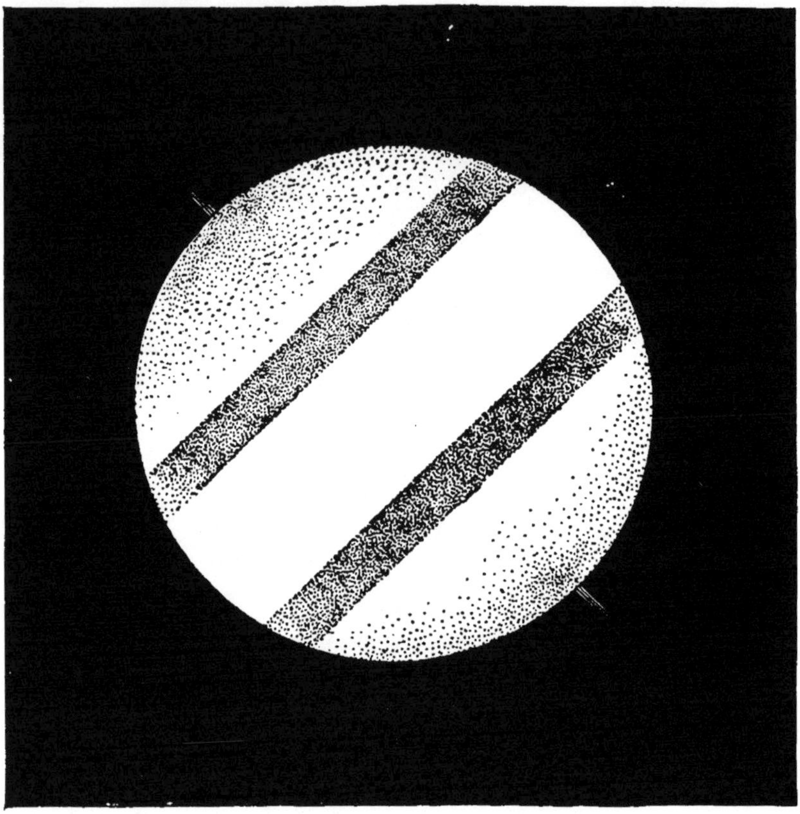

Fig. 9. URANUS. (Rings Forming.)

Uranus to-day seems to possess an uncompleted annular system. As the centuries roll by the astronomer may see the two dark bands near that planet's equator joined into one, and lifted into the Uranian skies, an actual ring-appendage, approximating at least the completed grandeur of the planet Saturn.

cede the "missing link." The germs that came in with the first ocean were adapted to the peculiar waters that fell at that time, and the germs that came in with the second downfall were adapted to the waters containing them, and the two kinds of germs were necessarily different, and hence the different forms afterwards developed from each. Hence the hiatus, hence the specific difference in forms with no "connecting link."

The geologist of the future in examining the infusorial beds of the earth, as well as the compound nature of some of the earth's giant prodigies, will, I am sure, welcome the annular theory to his aid. It seems to me that if all the upper vapors and their accompanying matter had descended to the earth in primitive times the constitution of the aqueous crust would reveal the fact by a very different structure. The reader can now see that the ring system, before it had taken the annular form consequent upon the cooling and condensation of the mass, must have contained some of the vapors of all the vaporized minerals and metals of the seething globe. That as the vapors cooled and contracted on the outer boundary of the mass first, but a small part of the heavy metals would be contained in the outermost ring and the greatest part would be contained in the innermost one. And that when the innermost ring fell to the earth a vast amount of the heavy mineral and metallic matter introduced into the vapors when in a heated or even superheated condition could not be held in solution when cold, as it could be when heated, and must therefore have been thrown down from the archæan waters as precipitates. The next ring would contain minerals and metals, but to a small extent, and the prevailing minerals and metals of the outer rings would be of a lighter character or in a lighter condition. Doubtless there would be traces of all minerals and all metals in all the rings even to the very outermost. So that every time an ocean of matter descended to the earth there would be some prevailing element in that downfall, and consequently one prevailing series of organisms. Any one can see that this would be the arrangement according to law. Now geologists must know that in each grand age this was the case; that in the earliest period of aqueous deposition the metals by all odds prevailed. In the silurian heavy calcareous matter prevailed. In the devonian the silicious and silicio-calcareous matter prevailed. In the carboniferous age carbonaceous matter was the characteristic element, and each ocean had its characteristic life-forms. Now suppose that all the oceans had reached the earth's surface in pre-laurentian times with all these elements held in solution and suspension. In this case it is plain that all these mineral and metallic substances would have been precipitated or deposited in and from the primeval ocean, and all subsequent formations would have to be derived from these early formed beds, or from cosmic space, and life would have been one continued evolution without a break. But how can we imagine that after 40,000 feet of the first formed beds had been laid down, and the oceans expurgated, we may say, the terrestrial waters the world round took up another vast fund of matter, first in the shape of the carbonate of lime (which it could not have obtained in such vast quantities from the laurentian beds), which being also deposited, the same waters after covering up a great part of the earth with a massive bed of lime, instead of re-charging themselves from this lime bed, refused the opportunity, and went thousands of feet below it, and took up an enormous load of magnesian lime and deposited it upon the carbonate (having called in the creative fiat to fill the ocean with new organisms). A well-known geologist, after having been staggered by these inconsistencies, concludes that "We seem compelled to ascribe the difference in the composition of the limestones to a vital rather than to a chemical or physical cause." With all due deference to high authority, I must ask, Is this question settled? Are we to expect men to make brick without dirt, or mortar without mud? By what imaginary process are the marine organisms, such as the millepores and other invertebrates, to build thousands of feet of magnesian lime rock, extending over millions of square miles, unless they are supplied with building materials? Whence did these lithophites get the magnesian lime with which they built this mighty casement of rock? This same authority says: "A preponderance of these or similar organisms *might produce a magnesian limestone.*" (Italics mine.) This question is not settled here. Neither a millepore, nor any number of them, nor any "organisms," "similar" or dissimilar, can make a lime rock of any kind unless that kind of material was at hand as a magazine of supply. Again the same geologist says, as if to share the responsibility with another: "Prof. J. D. Dana has shown that the millepore contains magnesia," etc.

Now I presume we all understand that these animals built the dolomites of the silurian out of the materials within their bodies! This, the vital process that built the magnesian lime beds? Thousands of feet of lime rock of continental extent, built by marine organisms out of their own bodies! But where did these animals get the magnesia that formed a "large percentage" of their bodies? Answer: From the ocean in which they lived! Here, then, we are forced back to our very starting point. We have an ocean strongly impregnated with magnesian lime, and yet no adequate source beneath the annular system. What would my readers think of my logic if I should attempt to show the source of this magnesia in the silurian ocean by putting countless millions of organisms in it? We do not want to know how the millepore secreted the lime rock, nor how he got his magnesia. We want to know how the ocean got it! Now we know very well that if this silurian ocean did not contain magnesian lime in solution the millepore would not have been there. If the waters of the ocean did not contain oxygen, where would we find the whale? Here is the logic of this question: My poultry-house contains a chicken-thief with a "large percentage" of fowls in his possession; this explains how my chickens got into the house? Moreover, if a "large per cent." of feathers should be precipitated on the floor we must conclude that the thief made them.

But leaving all humor for the pure administration of law we find a great downfall of annular waters with their accompanyiny life-germs,—a body of waters strongly impregnated with magnesian lime, and as a pure result the millepore and such other organisms as found here a natural habitat, came into being therein, and contained a "large percentage of magnesia" in their bodies because it was their home! The formation of the magnesian limestones came as a legitimate consequence of its presence in the water, so also came the organisms according to the demands of law. Such limestones as were not precipitated from actual solution, of course were formed by vital processes. No one will hesitate to admit that such a process will build a lime stratum. But unless lime is contained in the ocean as a solution the stratum cannot be formed; neither can the builder be there. Then since we are obliged to find a source competent to supply this measureless fund of lime, independently of the millepore, both carbonate and magnesian, we are compelled to bring in the aid of the annular system. Here the difficulty vanishes. For, as before shown, there was a down-rush of snows, a glacial period, an extermination of species, and a great uplift of strata just previous to the formation of this precipitated bed, which now comes in as a master-link of evidence, and which shows that there had been an addition to the ocean's volume at this very time, and which must have come from the annular system.

I presume, then, that this question is a legitimate one: Why was there a silurian age when the oceanic waters of the entire world were changed, as shown both by their fossils and the character of strata? All geologists know that the change from the eozoic to the silurian waters was a world-wide one, and how could it be possible for the same waters that gave origin to the Potsdam sandstone at the base of the silurian system, in America, and the kindred lingula flags of Europe, supply the stupendous world casement of lime unless there had been a vast augmentation of lime-waters? These lime beds are too vast and measureless to be made by the same process that obtains to-day. Now the ocean must have obtained its lime and its organisms together. It is extremely doubtful whether there were river systems in that early period, and consequently river erosion was a very small and unimportant and uncertain factor; and the evidence is daily accumulating that rains and fountains and streams were reduced to a minimum. It was not until the continents had been worn into corrugations (hills and valleys) that fountains could burst from the hillsides, and not until living fountains existed could there be perennial streams. How, then, did the infant continents supply detritus? By running streams? Surely not, unless these were fed by constant and excessive rains. But almost every evidence of the record points to the fact that the pre-existing continent whence it is claimed that the silurian seas obtained their lime was a very small affair. Thus it makes no difference how the geologist supports his claim that the lime beds of the silurian were derived from older continents, he is continually arrested by the demands of law. I presume the lime beds deposited in the silurian waters are vastly greater in volume than all the lime that has been carried from the existing continents.

But the difficulty does not stop here. The silurian lime was laid down. The stupendous piles of devonian rock were placed upon them. The carboniferous came and placed a heavy mass of rock of all kinds upon the devonian. The permian ocean rolled its waves over these, and left its load. Thus, for countless ages, the continents were worn away. Then the triassic flood deposited its load. Then the jurassic. The most of these were oceans peculiar and characteristic of their times. All had their peculiar organisms. As before shown they began their career in the midst of violence and change. Immediately after the jurassic period came the cretaceous, an ocean that washed the shores of the whole known world. Now geologists are well aware that the waters of the cretaceous period were radically different from those that preceded it. In many parts of the world cretaceous or chalk beds were deposited—calcareous beds radically different from any other lime-formation. From what continents did this ocean get its lime that it should be so different from every other lime-rock of the earth? As we have in this case, as in all others, the glacier and flood, crumpling and death, it is indeed fitting that a new deposit at the same time should predicate a new ocean. But a new ocean points to the waters on high; so does the glacier, so does the flood, the crumpling of crust, and the death of races. If the cretaceous deposits were only local we might attribute the change to local causes. But it was a change that left its way-marks around the circuit of the earth. It is these far-reaching and sometimes universal changes that direct us unerringly to the tellurio-cosmic matter of primitive times.* Before we close this chapter let us examine one more feature.

At the base of the silurian is a well-known formation in America known as the Potsdam sandstone, and in Europe as the lingula flag. This is claimed to be the product of the retiring waves after the archæan upheaval; or the advancing waves of an encroaching ocean. Were I to advance the claim that this entire

---

* The cretacious beds of chalk are made up of microscopic organisms, strikingly similar to those found in dust showers and colored snows, and could not have been derived from terrestrial beds. Hence, I see no escape from the conclusion that the chalk beds and their organisms were derived from the annular system. It would also seem that the flint and other concretionary forms therein had the same origin, and the claim that that system was the seed-bed of organism becomes well-grounded.

deposit was rather the wreck of rings than the wreck of continents, my readers might think it was rather the wreck or spoils of intellect. But look at the vastness of this formation. It underlies the silurian of the world. It spreads from Canada to Texas, and from the Alleghanies to the Rocky Mountains, and for all we know it is a casement surrounding the world. Waves must have had this stupendous wreck to work upon, or they could never, so I think, have laid such a general bed over the continents. The whole mass shows pre-eminently a mechanical and rapid accumulation. All that is required for this wonderful outspread of silicious beds over the continents is a telluric ring of silicious matter, in which was an opportunity, nowhere else afforded, for granulation and the growth of concretionary structures and infusorial germs. Indeed, if the countless millions of aerolites that reach the earth's surface can form in interplanetary and interstellar space into solid stones and pebble-like forms, what reason can be shown that a ring of vapors surrounding a world once melted and glowing,—a world in whose inveterate fires silicious matter was vaporized and made to commingle with the vapors of the ring system,—I say what reason can oppose the idea that these silicious vapors did not strongly impregnate the aqueous vapors and call in its life-germs at that time since they were both in the system together? But if in the system together they must have separated and segregated as the system condensed into granular, crystalline and concretionary forms, the very materials in the stupendous outspread at the base of the silurian system.

Let it be distinctly understood that the annular theory admits the universal eroding powers of rivers and waves; the transportation power of currents and the continental process of strata-building from detrital matter. But with all this admitted, the wave of ocean can do nothing in this line of work unless it is supplied with matter with which to work, and the simple question is, Where did the wave get this crystalline and granulated and infusorial matter to spread over the floor of the silurian ocean? An ocean so far as the accompanying strata show was spread over the American continent at the time of its deposition. It underlies, so far as our information shows, the whole continent of North America. Now it is altogether out of reason to suppose that this wonderful mass is a detrital formation worn from surrounding continents unless we admit that every foot of it was a playground of shore waves, either advancing or receding. But we cannot find that there was, either before or after, a continent here surrounded by such an ocean. We can, however, find a fund of silicious matter all adequate and necessary in world-making,—a fund that existed before a wave ever washed the earth, and which must be found in the earth's crust. When it fell with the descending waters we can conceive how naturally it would form a world-wide bed of matter. I presume no man who is led to admit the former existence of a great primitive aqueous envelope surrounding the earth will for a moment hesitate to admit that those primitive vapors did contain calcareous, silicious, carbonaceous and metalliferous regions consistent and favorable to some forms of life-germs, and that when the ring segments fell to the earth these very substances spread over the ocean's floor as individual beds. We have seen the great metallic beds in the foundation rocks of the continents just where we would expect to find them according to annular arrangement with primordial forms. We have found great primitive calcareous beds in harmony with this order; and when we take a comprehensive glance at the Potsdam beds and their kindred and co-temporaneous beds under other names, spreading as a coarse mechanical deposit over so much of the known world, what philosophy can be urged in support of the claim that all parts of this wide expanse have successively been the ocean's playground of shore-waves? Thus we find that our theory leads to astounding conclusions. The ocean's waters have built numberless beds of limestone through the instrumentality of marine organisms, but the annular waters supplied the builders with materials and set them to work. Vast beds of metals have been laid down from the ocean's waters as regular stratified deposits, which could never have been borne from archæan terranes, and must therefore have been supplied from the only other source,—the primitive vapors that we know contained them. All over the world, in all times subsequent to the archæan, great sand beds have been formed, and these originally came from the annular fund. But let it be remembered that while this must have been the original source of these materials, the ocean's waves have never ceased to work upon the wreck of rings, devouring the original beds and transferring them to other lands. Rivers wherever they ran have eroded with resistless appetite, and fed the ocean with its own beds formed again and again. I freely admit that this work has progressed from that early day when the first ocean baptized the infant earth; but in addition to all this great labor of tearing down and rebuilding, a hundred times or more the annular waters have descended, and in their rush to the seas have laid mechanical beds of materials torn from millions of valleys, which could never have been otherwise formed, and at the same time added their own fund of exotic matter to these beds. Exotic matter of all kinds—lime, iron, sulphur, salt, etc., etc.—have hereby entered into the various formations of the later ages. We have seen in almost all lands great beds of iron in the mechanical deposits of later ages. In the hills of my own State and country are stupendous masses of iron ore—true sedimentary formations, more than 400 miles distant from any primitive metallic beds. Tell me, how was this metallic matter carried from older beds and deposited here? If I am told that it was the product of organic distillation I must demur, and the rocks themselves are my witnesses. If told that the annular system was permeated by metallic dust, and that as it reached the earth in almost all lands beyond the tropics, we can readily understand why there is so much metallic ore of lighter specific gravity in the more recent geologic strata; and when I turn and see these beds more abundant in regions outside of the tropics, where upper matter must have fallen, if it ever fell, and when I reflect that if these ores had been a vegetable product they ought to be more abundant within the tropics where vegetation is most abundant, I am led to conclude that much of the iron ores of the carboniferous beds had an annular origin.

Suppose a grand decline of annular matter from near the middle of the system should occur. Where would we expect to see its effects? If they were registered in the rocky volume, would we not expect to find them in that geological horizon embracing the later devonian, the carboniferous, etc., perhaps extending into the cretaceous? Since then we would expect to find the very heaviest matter in the innermost rings, and the very lightest in the outermost, and having found this heavy matter in the first formed beds, and the lightest prevailing in the latest, we must expect to find this mid-horizon characterized by a more abundant supply of matter of medium specific gravity from the middle of the annular system. We certainly would not expect to find the salts of soda and lime largely developed in the same annular region with iron, lead, silver, copper, etc., nor on the other hand would we expect to find these salts to much extent developed in the latest formed beds. The last descending vapors must have been nearly free from these. But where in the geologic column do we find such substances more largely deposited? In this mid-empire of the ages! Where are the salt and gypsum formations of the world? Even those on the very surface of the earth, as in the Rocky Mountain region, have been derived from beds of these middle ages. I am sure a full history of these medial formations, when written, will present this philosophic order of original deposition, and that all more recent beds may be traced to these first deposits. Enough is already known to show this order; this intelligent plan laid in the earth's primitive envelope. Can we at most show any reasonable method by which the oceans of medial ages obtained these salts from the primitive beds, even if they were known to contain them?

And now when we take into consideration the reliable statements of Arctic explorers of the existence of salt beds, salt marshes, etc., in polar lands, where, above all, we would expect to find the least, according to the old theory, and the greatest quantities, according to the new, what can be urged against the claims here set forth? These are the considerations that will bring the annular theory to the test. To these I freely submit it.

Thus the annular system, the great seed-bed of organisms, in its final wreck and new arrangement in the super-crust, not only becomes a self-supporting argument of an intelligent plan in strata-building, but it affords the key that unlocks the deepest mysteries of organic evolution in the measureless ages of the past. If men of this age refuse to use this key, other men will gladly embrace the opportunity when we are in our graves. My chief desire is that some persons more competent than I am may take these things here presented in their rude state, and prepare them for the majestic building which the faultless Architect has planned, and who is calling for builders.

Can these be overdrawn conclusions? What do the great "gaps" and "missing links" in the record mean —these evolutions of life-forms brought up to a certain limit, then a cessation, then a leap forward on the plane of progress? If we admit that each ring-section uecessarily held its own peculiar life-germs, and each succeeding one germs one step nearer perfection in the goal of life, as demanded by their position in the system, how the confusion vanishes! How harmonious the gradation, from the moneron of the first or primitive life element in the dark deep of chaos, to the grand platform of physical light and life! If we spurn these conclusions, where do we stand? Spurn them, and we are compelled to admit either numberless specific and especial creations, or a most unreasonable and unphilosophic evolution of one species from another. Reject these conclusions, and we will then be forced to admit that all the multifarious forms of life now on the earth, including man, were planted potentially in organisms of the archæan waters; and further, that this plan of evolution was so directed by the Creative Hand that it stopped short again and again, and started again upon a loftier platform of life—i.e., with new forms, only to stop and leap again. Which line of evolution will the reader choose: That which carries the whale from the rhizopod of the primitive seas over numerous breaks in the line, or that which leads it to perfection from its own original life-germ planted in its own soil? Is it not more reasonable, and in the line of law, that man should arise from his own especial organisms in his own environment, a germ of God's right-hand planting, than that he should descend from the little moneron of the eozoic seas by innumerable sudden starts through countless millions of years in the specific organisms of a thousand environments, as the oyster, the ammonite, the fish, the bird, the ape, etc., etc., to the illustrious Darwin? I am not willing to admit that that great man, to whose memory the world owes a debt it cannot pay, came through such a line of descent when a more philosophic and intelligent one is open to my view.

CHAPTER XIII.

A CONSIDERATION OF THE EVIDENCE THAT INEVITABLY LEADS TO THE CONCLUSION THAT THE CARBON STRATA OF THE WORLD WERE DEPOSITED AS AN AQUEOUS SEDIMENT FROM THE EARTH'S ANNULAR SYSTEM WHERE IT HAD REMAINED FOR COUNTLESS AGES AS A PRIMITIVE DISTILLATION, EXPELLED FROM THE INCANDESCENT OR BURNING EARTH.

In millions of bogs and swamp marshes in the northern part of the Northern Hemisphere the decay of a characteristic vegetation gives rise to carbonaceous beds of matter called peat. This peat when placed in the retort of the gas furnace, and subjected to heat, as in the manufacture of burning gas, may be made to give rise to various products or distillations, from the heavy form of graphite and asphaltum to the lighter forms of oily hydro-carbons, such as are found in the earth's sedimentary crust, under various names, and chiefly as stone coal, or simply coal. It was very natural for the philosopher to conclude that the coal strata of the earth were mineralized or metamorphosed vegetation, since it was well-known that peat, subjected to the proper treatment, might be made to yield these products. And as there is now no other terrestrial means than rock pressure, and the native heat of the planet's crust, adequate to produce these products, men were honestly led to the conclusion, which now prevails, that the coal beds of the world are of vegetable origin. It will thus be seen that I admit—that all men must admit—that vegetation, when the necessary conditions are present, must become a mineralized carbon fuel. To conclude otherwise would be a fraction of inexorable law. Hence no future criticism upon the theory I shall presently advance would be in order upon this branch of the question, since I am forced to stand upon the self-same foundation that all scientists must stand upon.

But standing upon this foundation we must erect an edifice in harmony therewith. We must not allow a stone to enter the structure that has not been squared and dressed by the Master Hand of Law. Stones, lying ready prepared in the vast quarry of nature, must be our building material.

Then with the full understanding that the slow combustion in a swamp marsh or peat bog under favorable conditions gives rise to fuel carbon of various degrees of mineralization, we will begin our examination of this momentous problem.

First: It will be fully conceded by every scientific and philosophic mind that the natural change of vegetable organisms to the form of elementary products in peat formation is a combustion, or slow distillation, by which the elements of the compounds forming the body of the plant are dissociated and made to pass into other forms. And in which combustion carbon particles released from their associations remain in an unburnt condition. In other words, the carbon comprising the peat bed is simply unconsumed carbon.

Second: This unconsumed carbon product of swamp combustion is the same as the unconsumed carbon of any other combustion or distillation in which the carbon element is involved, under whatever circumstances said combustion takes place.

Fig. 10. SATURN (RINGS FORMED).

These are, as all can see, self-evident propositions—propositions to which we are all irretrievably committed, and I want the reader to see that we diverge not from this in our line of argument. I ask my brother geologists to give me their attention for one hour; and I will give them in return for their kindness a theory of coal formation planted upon this rock. A theory that must be true from the very nature of the problem, and which, if true, must explain every difficulty involved therein.

From these considerations it is manifestly certain that the combustion that takes place in a stove or fire-place is precisely similar, except in intensity, to that which evolves unconsumed carbon from swamp vegetation, and that the unconsumed carbon that arises from our chimneys and locomotives in the form of smoke is the same carbon element which, resulting from the decay of peat vegetation, sinks to the bottom of the marsh. The processes are necessarily parallel, producing the same elementary changes and the same products.

One process deposits its unburnt carbon in the bog. where it is sealed away from that universal devourer, oxygen, and where it remains a veritable fuel. The other process sends its unburnt carbon into the air,—into an ocean of oxygen, a veritable fuel, which is immediately re-burnt and converted into invisible carbonic anhydride by its union with this free oxygen of the air. If no oxygen were in the air ready to seize upon this carbon fuel, the atmosphere would in a short time become filled with it, and as it became saturated with the moisture with which it comes in contact it would settle upon the earth as carbon-dust—a veritable fuel, the very same in kind that is sealed in the peat bog. Some portion of the unburnt carbon which arises in our fireplaces does actually escape this devourer, and adheres to the back wall, chimney or stove-pipe as an unconsumed carbon product—a veritable fuel, which takes fire and burns as every one will admit. Who has not seen the "chimney on fire?" Who has not seen the oily carbon on the back wall where wood or bituminous coal is burnt take fire and burn—thus proving to every beholder the unavoidable conclusion that the carbon or smoke that arises from every chimney and furnace of the earth when measurably shut up from immediate union with oxygen remains an unburnt fuel, precisely the same in kind as the unburnt carbon fuel of the peat bog?

If we were to collect the unburnt carbon from our chimneys into piles, where moisture and air could have free access, it would take fire spontaneously and burn; as it is well known that many a disastrous fire has occurred from this source alone; just as peat dug from the bog sometimes takes fire spontaneously.

Thus in every particular the smoke that arises from every combustion in which carbon is an element is an exact counterpart of the carbon arising from the decay of swamp or peat vegetation, and hence we are compelled to accept this conclusion: that the millions of fires, foundries and volcanoes of the globe that pour immeasurable volumes of unconsumed carbon into the atmosphere are forming fuel wherever soot is formed; and that if it were not for the ever-active oxygen of the air it would all descend upon the earth as a fuel, and become incorporated in the forming sedimentary beds of the earth as such, and under favorable circumstances it would be collected by water currents into beds of fuel! Here, then, just as we are entering the threshold of the coal question, while starting committed to the fact of peat formation, we are driven by the implacable demands of law to the conclusion that there is another and a parallel process producing the very same effects! How can we possibly avoid the conclusion, then, that if a terrestrial fund of rising smoke should be in sufficient quantities, and should arise, amidst a fund of aqueous vapors beyond the reach of the devourer oxygen, that all the coal beds of the earth could have been formed by that returning fund? There is no avenue of escape from this conclusion!

Our next duty, then, is apparent. If the coal beds of the world were not formed by a process of vegetable decay, they were, in whole or in part, formed by this parallel process! And we must now proceed to show that although a fuel carbon is necessarily formed when vegetable decay is arrested in a swamp marsh, yet this decay or combustion is so complete that the products are utterly inadequate to form great continental coal strata. When we shall have shown this we will proceed to show that an immeasurable fund of unconsumed carbon, or simply smoke, went up from the igneous earth for countless ages, pouring a vast fund of fuel carbon into the primitive vapors that surrounded the burning and incandescent sphere and which fell with those vapors from our annular system, and floating away into the ocean settled upon its floor, with the vegetation involved.

Men to-day are in the habit of pointing to peat bog distillation of carbon as the origin of coal in its earliest formation. This is only one more illustration of the universal disposition of the human mind to accept conelusions drawn immediately from appearances only. Such illusory evidence has more than once involved some of the sublimest truths in the deepest clouds.

When the plant dies and begins to decay one of its constituent elements, carbon, oxidizes in a process of slow combustion, and returns to the air as an invisible gas. Now it is only when as by accident a particle of this carbon fails to become oxidized that it remains as an unconsumed atom, and then by accident becomes sealed away from oxygen, ever alert and active. It is therefore an exceedingly small part of the world's vegetation that is left as unburnt fuel in peat bogs. Let us not forget that this swamp combustion is precisely that which takes place in every furnace and fire-place on earth (time being left out of the account). Just as a very small part of the smoke ascending from the coke oven or volcano is left and formed into a fuel, and fails to combine with the oxygen of the air, and consequently adds but an infinitesimal amount to the unconsumed carbon of the earth, so is it with vegetable decay. Now we all will admit that all the carbon in the earth's crust was derived in a primitive distillation from the mineral world that originally contained it. Were it not for this primitive process an atom of peat carbon could never have been found; for the plant could never have obtained it. The formation of peat carbon, then, is at most a secondary process; and who will fail to see that if through vegetation and peat alone carbon or coal had its origin, then the primitive process is entirely ignored? But every man of reason must own that the very primitive process that gave carbon to the primitive atmosphere for the use of the plant gave it forth in the same form that the plant itself does

—as an unconsumed carbon fuel! Why, then, should the earth in after times institute a secondary process to produce the fuel form of carbon it already had formed by the primitive process? Then I repeat that on the very threshold of the coal problem we find that we are forced by unyielding law to admit that there was a stupendous fund of fuel carbon produced by a process parallel to peat distillation, but previous to it in time. In short, there is no way of escaping the conclusion that every atom of carbon in the coal beds of the earth, even if they were wholly a vegetable product, was previously produced from the mineral earth by an original process. Consequently the discussion of this problem in the very beginning demands serious consideration from men of thought.

It cannot but be then that a secondary and fortuitous process of fuel making must fall behind the original one in importance. As in the great supply of lime to the primitive ocean there was a call and demand for organisms to use up the surplus of calcareous matter in aqueous solution after the beds were precipitated, so after the deposit of the fuel carbon was there a demand from the earliest ages for vegetation to use the surplus carbon in the atmosphere. But what a puny process compared with that grand sublimation and distillation in the igneous earth!

But let us admit that by some unseen and fortuitous means the original carbon fuel was afterwards metamorphosed into plant food, and was eventually retransformed into fuel. Let us admit this just to see how we can come to the erudite conclusion that coal is a vegetable product. On this supposition every atom of carbon in the carbon-beds of the world has existed in the air, or elsewhere, in union with oxygen as an invisible gas (carbonic anhydride). For in this form alone can it enter into the economy of the plant. But to insure this production of plant food the carbon, as it issued from the fires of the igneous earth, must have been poured into an ocean of free oxygen. For if this were not the case the carbon still remained an unburnt fuel, which the plant could not use. Now while it is very unphilosophic to suppose an ocean of free oxygen attending an igneous or incandescent world, it is also susceptible of the plainest proof that if all the oxygen in the super-crust had been present in the primitive atmosphere it could not have saturated the carbon of the coal measures. Then, again, if this remaining unburnt carbon ever afterwards became plant food, in order to produce coal, it did so through the process of spontaneous combustion, after it had been once formed in the presence of oxygen and sealed away from its ravages. But if this be true, how does it happen that the very same unburnt carbon, in the form of peat, having the same opportunity to spontaneously burn, does not also undergo the same change? The conclusion, then, is inevitable that unburnt carbon fuel of the igneous era was stored away in the earth's crust. And if not still there has suffered spontaneous combustion, and has been again converted into fuel, which cannot spontaneously burn, which is simply preposterous and absurd. The fact that the carbon in the peat beds of the earth has not thus disappeared is substantial evidence that the combustible unconsumed carbon of the igneous era did not, since they were necessarily the same in kind. Now where are we? The advocate of the vegetable origin of coal is compelled to hang upon one horn of this dilemma, or both, if he choose. If he advocate the existence of a universal sea of free oxygen around the igneous earth then he must admit that there were no residual carbon products, and we will grant this conclusion, for argument's sake.

No residual carbon products mean no primitive carbon beds, and that all coal beds are mineralized vegetation. But these necessitate an ocean of oxygen. Let us grant this. But this forces us to the conclusion that immediately after the igneous era and during all the ages from the archæan to the later tertiary there existed about the earth a universal sea of plant food (carbonic anhydride). Dana, seeing this inevitable conclusion, says: "The atmosphere now contains less carbonic acid than it did at the beginning of the carboniferous period by the amount stored away in the coal of the globe," and yet the same high authority says "much more carbonic acid [than now exists in the air] would be injurious to animal life." *

I must take a little time to examine this, remembering that even during the carboniferous era, air breathing and water breathing animals existed in abundance;† remembering that water is a great absorbent of carbonic acid, and that "much more carbonic acid" than exists now in the atmosphere would kill every mammal, fish, bird or salamander on the earth.

Admitting that every foot of the earth's surface supports an amount of oxygen equal in weight to at least 420 pounds, as the best authorities teach, and allowing a cubic foot of coal to weigh 70 pounds, considerably below the average, and the average of charcoals as 10

* Read Dana's "Manual" from page 340 to 353.

† See Dawson's "Acadian Geol."

pounds per cubic foot, and taking the conclusion of chemists that a bushel of charcoal will yield 2,500 gallons of carbonic acid, we find one cubic foot or ten pounds of charcoal will yield a little more than 2,000 gallons. Now a cubic foot of coal yields from 40 to 60 pounds of carbon; we will put the average at 50 pounds of pure carbon. Then it is plain that if 10 pounds, or one cubic foot of charcoal, or nearly pure carbon, will yield 2,000 gallons of carbonic acid, that 50 pounds will yield five times as much, or ten thousand gallons, and a column ten feet high and one foot square will yield ten times as much, or 100,000 gallons. Then calling eight gallons equal to one cubic foot, it would make a column of carbonic acid 12,500 feet high and one foot square. Here, then, is revealed the astounding fact that if all the coal of the earth (including the graphitic coals of the archæan, etc.) combined would make a bed of pure carbon 10 feet thick around the earth, it actually drew from the atmosphere an ocean of carbonic acid 12,500 feet deep extending around the entire sphere. Now if my hypothetic vein of 10 feet be not an exaggeration, this is at least the amount of carbonic acid the carboniferous atmosphere contained more than it now does.

According to Youman there is enough carbonic acid in our atmosphere to make an ocean only 13 feet deep. And eminent physiologists say that three or four per cent. of the present atmosphere in the shape of carbonic acid would be fatal to life.

But is my estimate of the hypothetic coal vein of 10 feet too high? Is there enough carbon in the coal beds to make a world stratum ten feet thick? Late discoveries of coal in almost all lands outside of the torrid zone would induce me rather to increase than diminish the estimate. Taking the vast coal fields of the Mississippi Valley, Texas, and the great region of the Cordilleras, I presume one-fourth part of the United States is underlain with coal veins varying from 1 foot to 25 feet. In Pennsylvania it will aggregate over 40 feet in thickness. In Eastern Ohio it will reach almost the same thickness. In Eastern Pennsylvania a single vein sometimes attains the thickness of 30 feet, with from 7 to 10 massive veins. In Nova Scotia there are more than 70 veins; one of these is 38 feet thick, another 15, and another 12. In Great Britain there are at least 100 coal veins, amounting, it is said, to a thickness of 300 feet. In France are coal veins 100 and 120 feet thick. When we turn to the vast coal fields of Asia and Southern Africa, South America and Australia, Alaska and Greenland, and then count in the calculation the vast beds of carboniferous shales and oil-bearing strata of carboniferous limestone, and remember that Dr. Dawson has said that there is likely as much carbon in the archæan rocks as in any subsequent formation, and that these beds are world-wide, I think we need not diminish our 10 foot seam.

But lest the reader may think that I am claiming too much I will diminish this stratum to five feet, and we will yet have an ocean of carbonic anhydride more than six thousand feet deep. Or if I am compelled to reduce it to one-half this much we will have more than 3,000 feet, or three times as much as Youman's four per cent., or two hundred and forty times as much carbonic acid as now exists in the air. And yet Dana's fishes and amphibians survived!!

No wonder that a modern chemist, investigating

these measureless carbon beds, came to the conclusion that there is not enough oxygen in the air and crust of the earth combined to saturate the carbon of the coal veins without the deoxidation of its salts and ores.*

Hence it is plain that there could not have been such an ocean of plant food. The existence of breathing animals in the very midst of the coal-forming period forbids it: The simple fact that at the very time the primitive carbon was being distilled in a burning world the oxygen under chemical law had a choice in combining with other elements present in the great alembic, forbids it. We are, therefore, left without the slightest chance to oppose the claim that the primitive fires in the molten earth did produce measureless quantities of fuel carbon, and which has always remained such. Where is it?

Now as the formation of fuel in a swamp is the same as that in every fire-place, foundry, furnace and solfatara on earth, and both secondary processes producing secondary products, and as the latter, as all must admit, is utterly powerless to add materially to the fuel forms of the earth as coal, so must the former be powerless to add greatly to coal formation.

If there be so much as one feature of the coal problem which the primitive carbon theory fails to explain after a fair test, then it must be a failure. It is then with the utmost confidence that I prepare it, knowing that according to eternal law we shall find beds of primitive fuel carbon in the earth. I will now in as brief a manner as possible specify some conditions that must prevail in the coal beds as decisive tests:

First: As the annular system was without doubt a

---

* Phin's "Six Days of Creation," page 66.

region of microscopic life and infusorial forms, I presume a coal bed must be largely characterized throughout its mass by the presence of microscopic organisms. While at the same time, being a deposit from sea waters, it must have carried down organic forms existing in those waters, and remaining for a while lying upon the sea bottom before it became covered up by other beds, it must have become as other oceanic ooze, more characteristically marine upon its surface (in marine waters) than in other parts of its bed.

Second: These carbon sediments must have borne down a vast amount of marine vegetation, and buried it upon the sea bottom, and must also have accumulated in beds on the land surface, but here only in the lowest region—i.e., in swamp marshes—and here the involved vegetation would be different, the marine character being largely absent.

Third: Where a carbon-fall was borne to the seas, that part of it which settled where limestone strata prevail would indicate great distance from the shore, and here the roof shales of the coal seams must be measurably free from land fossils. While coal beds among intercalated sand strata would indicate deposits nearer shore, and here the roof-covering of the coal beds would likely be more truly mechanical beds, or at least contain land fossils to a more liberal extent.

Fourth: As all downfalls from the annular system must take place more largely in regions distant from the equator, the coal beds must be more heavily developed toward the polar regions than elsewhere. Those located nearest the poles must be the nearest free from terrestrial or marine impurities, and yet with such impurities eliminated must be specifically heavier; while those floating farthest from the region of downfall toward the equator, or into bays and gulfs of the ocean, would contain a greater amount of impurities or ash, and with the ash eliminated would be specifically lighter.

Fifth: All carbon downfalls must have been attended by great cataclysms of snow or water, or both, and more likely than otherwise the periods of coal accumulations were those essentially indicative of violence, if not of cold.

Sixth: While a bituminous coal vein deposited in regions subject to volcanic strains and mechanical heat arising therefrom, would necessarily be metamorphosed into heavier and harder carbon forms, as into anthracite, etc., yet as there must have been all the light and heavy forms of carbon in the annular system, as primitive distillates, it is certain that all these forms of carbon may be found in lands where no strata disturbance has taken place.

Seventh: When a carbon-fall took place, and the carbon was borne to the deep seas, the heavy carbon such as the anthracite and semi-bituminous particles would settle in the deep ocean, while the lighter, not being able to reach bottom, would float to shallower waters and settle as lighter coals, and according to this view a submarine valley might have a deposit of anthracite carbon, while a neighboring bed on an elevation might be a bituminous deposit.

Eighth: These facts must lead us to the inevitable conclusion that in both the Northern and Southern Hemispheres the coals must be more valuable as we proceed from the equator and the least valuable coals must, as a rule, be nearest the equator, and also in smaller quantities. I will offer this as a decisive test, if the reader choose.

Ninth: As there must have been carbon disseminated throughout the annular system there must have been carbon-falls in all ages! And the earliest falls necessarily the heaviest and purest, and the last falls of carbon must have been the lightest and of the poorest quality; and if any downfalls from the annular system have occurred in recent times then this light carbon must be found on the very surface of the earth, and imbedded in the snows that fell with them in the polar regions, and must also form the foundation of recent peat formations in cooler regions.

Now some of these points I have not had time or opportunity to investigate fully. But I propose them with the utmost confidence, for all must see that they are legitimate conclusions, and men of science can by these confirm the theory or hurl them with deadly effect against it if they are not true. They are all important and decisive tests. The ninth or last embraces a volume in itself, and I regret that I have so little space in this to consider it. I have elsewhere referred to the vast reaches of carbonite spread over so much of the Northern Hemisphere, but it is important that the peat bog question should be settled before we venture far into the coal problem. We will attempt to settle it now.

Why is this carbonite covering confined so exclusively to regions glaciated and submerged during the last reign of ice and flood? To go back a little in order of time, why are the terraces that were built up by the flood-waters of the retiring glaciers in the unglaciated valleys so frequently characterized by the presence of carbonite called "peat," "lignite," etc.? Is it not a demonstrable fact, for instance, that the great terrace beds of the Ohio and Mississippi Rivers were carried by floods from the glaciated regions of the Great Central Basin? And are we not compelled to admit that the so-called lignite seams or peat bands in those terraces were also borne from the same basin? Then they must have come from the melting glaciers or fund of snows! These carbonite streaks, plainly visible in the Mississippi embankments from Vicksburg to the mouth of the Ohio, and planted deeply beneath the waters of the lower Mississippi at New Orleans, were formed as the flood deposits were formed, and must have been borne by the same waters that carried the body of the terraces themselves.

I have shown how this vast spread of carbonite was the result of the last great debacle of snow and floods. It is evident that if this black carbonaceous deposit were covered up, as the lignites and peat of the valleys of Europe, Asia and America are, it would be called lignite or peat, and no man would question the inference. But it does lie at the bottom of thousands of lakes and ponds. It has been dredged from the waters of the northern oceans, and the dredge will surely bring it from the Great Lakes, from Hudson Bay, and from the Arctic Ocean.

But in thousands, nay, millions of these ponds and swamps grows a so-called peat vegetation. And this is the question we must now endeavor to solve. Peat-bog vegetation or moss, known by the generic name of Sphagnum, is characteristic of the swamp, and grows only where peat is forming—a circumstance which alone should teach us to look back beyond the era of vegetation for its origin. It is plain that if no peat (carbonite) had originally and previously been deposited where the plant now grows, it would never have grown there. As the millepore and its kindred organisms would never have lived and flourished in the seas if lime had not previously existed there as the food, so neither could any of the sphagnous mosses have planted themselves over the medial and colder latitudes of the earth, if the carbon beds necessary to sustain them had not been previously planted there. As we are forced to look beyond the era of oceanic organisms into the annular system for a primitive supply of lime, so we must also look beyond the plant through the igneous and smoking world into the earth's primitive envelope for the food that called the peat vegetation into existence.

Now as we simply know that unconsumed carbon did exist in the annular system, and that its lightest forms were the last to descend upon the earth, not in the equatorial regions, but nearer the polar world, and since we find such an enormous outspread of such carbon not in the tropics but in the colder regions, accompanied by the plant demanded by its existence, how can we avoid the philosophic conclusiou that if there had not been a downfall of carbonite in the very last geological epoch, the sphagnous vegetation would not now exist there?

As the marine organisms came into existence after their food was supplied to the seas, and began their subsequent work of rock making, and are thus employed to-day, so the peat vegetation came and began its offices of peat making after its food was supplied, and is thus employed to-day. Then it must be apparent to every man of reason that peat vegetation now forming a secondary product of carbon is our first and unimpeachable witness of the primitive and annular origin of coal. There is no reason why peat vegetation should not grow more abundantly in the tropics than in colder regions, if the carbonite or sphagnum food existed there. There is every reason for believing that if coal is a vegetable product it should more abundantly exist in the equatorial regions. Does not the vegetation theory demand this? Why, then, are the great carbon formations planted where the annular theory demands that they should be found, and not where the vegetarian wants to find them? Is it not plain that if such beds of coal existed under the equator as are found in zones of eternal ice, that our theory could not be supported thereby? The presence of every coal bed in the polar world is an eloquent testimony against the current theory. The coal was planted where it affords the vegetarian no consolation. But with what an air of triumph could he point to it, if it were planted within the tropics, the very home of vegetation!

If we could by any means change the character of the bed of peat, the vegetation would languish and die, and another species of peat plant would succeed. This is abundantly proven by the fact that where lime waters or marine waters saturate the peat-bed, other species of sphagnum flourish and form other kinds of carbon. Thus we see that law leads us directly away from the current theory. The law that guides the acaleph of the sea guides the sphagnum of the bog. Each had its food supplied before it flourished in its own habitat. Each is to-day continuing the process of strata-building. But we must call it a secondary and

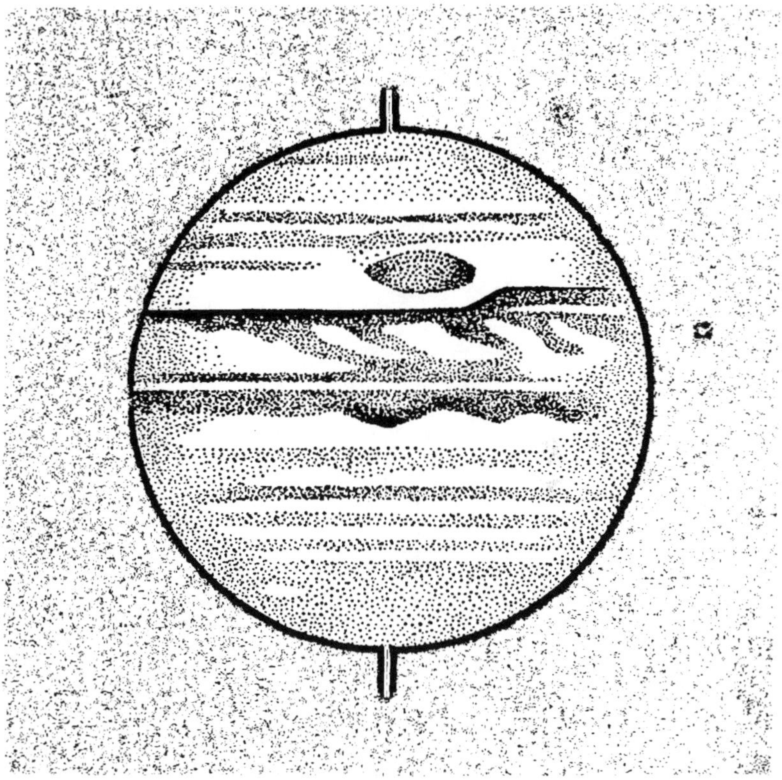

Fig. 11. JUPITER. (RINGS FALLEN.)

Jupiter, the King of Planets, is very likely an inhabited world. But what must be the canopy forms that are to-day directing thought and intellect on such a world? The face of Jupiter here presented was seen several years ago, but here are some forms that we cannot fail to recognize in the legendary annals of earth. I can only allude to *Leda* and the *egg*, which, under the power of Jove, the true sky, brought forth the "twins," *day* and *night* (Castor and Pollux), a problem that Max Muller well-nigh solved.

puny work, indeed, compared with the original and primitive one.

If a colony of ants build a mound a foot high, must we conclude that Mont Blanc is an ant hill?

If the hickory, the ash, the pine and the lycopod must each have its peculiar foundation soil, previously laid down, before it can take root and flourish, so must the sphagnum and the hypnum of the marsh. The pipsissiwa must have its shade; the epiphegus must have its beech tree, or it will not grow, and the bog moss must have its carbon bed as surely as it must have its air before it could begin its offices. And further, the succession of species and tribes of plants in the geologic ages as clearly demands a succession of downfalls of annular matter as the other witness in the record.

Having then, as I think, planted the primitive carbon problem upon the rock of philosophic law, or rather having found it planted thereon, we will next examine the *ipse dixit* of the coal beds themselves, and note the inevitable harmony.

CHAPTER XIV.

IS COAL A VEGETABLE PRODUCT?

AN EXAMINATION OF THE COAL BEDS UNDER THE LIGHT OF THE ANNULAR THEORY.

The accepted theory of the origin and formation of coal is simply this: It was formed of the remains of an ancient vegetation that grew largely in peat and swamp marshes. The adoption of this view has led to the inevitable conclusion: 1st. That coal is vegetable carbon changed to a hydro-carbon, and subsequently partially changed to an oxidized hydro-carbon. 2d. Each coal seam, however vast and boundless its extent, was universally submerged beneath the sea to receive its superposed beds of sand, clay and lime, and afterwards re-elevated essentially to the ocean's level to receive the next coal seam placed above it.

These points so warmly maintained by the great fraternity of geologists we will now examine by the piercing light of philosophy.

First let us remember that every atom of the great mass of carbon now forming the coal deposits of the world must have been a distilled product of a primitive igneous process, even before the plant could possibly appropriate it; and that we are forced to admit that this primitive and original process, which took place long ages before a plant ever existed, must have supplied the very same chemical products now found in the coal beds.

Every philosophic chemist is thus unavoidably

bound to the conclusion that a process a thousand-fold more stupendous and competent to produce all the forms of carbon now found in the crust of the earth existed in the great telluric gas furnace of primitive times.

Let us imagine two worlds, one covered with a swamp vegetation, amid which a slow and puny distillation is giving rise to the accumulation of carbon fuel; the other a boiling, burning and smoking planet, distilling and subliming from millions of furnaces carbon in all its forms. Which of these imaginary worlds would form fuel carbon more rapidly? Which would, for the time occupied and the means employed, be the more competent agent in the grand process of strata-building? If, in the primitive world-furnace, the unconsumed carbon fuel should form in the least degree, it would return in after-times, with its aqueous and other matter; and the mind is utterly at a loss to find figures wherewith to multiply the vegetative process to make it at all comparable with the igneous.

On the other hand we see the great gas retort of the molten earth, distilling every carbon product, from the heavy graphite of the archæan rocks to the light carbonite of recent times, and we know this process did obtain, if the earth ever was in a molten state. On the other hand, a process which we know could never, and would never have obtained, if the foundation had not been precisely laid and the food elements previously supplied by the former. On the one hand we see a process commensurate with creative effort; on the other, a process belonging to a world in its completeness. On one hand, we see, in short, a fuel carbon formed in immeasurable quantities, gathered from the only possible terrestrial source, and stored away as an aqueous sediment in the earth's crust; on the other, we see this ready-formed carbon as a fuel, etc., entirely ignored and disregarded in spite of law, and made by some mysterious process into plant food,—an invisible gas; a non-supporter of animal life; a non-supporter of combustion,—and again transformed into a solid carbon, and then by a secondary combustion transformed accidentally into a fuel oxyhydro-carbon, the very same thing previously formed. Looking back upon these hypothetic worlds as we start upon our tour of investigation, who will not say that the primitive carbon theory has a foundation a thousand times more permanently planted? Why force the puny process of peat formation to supply carbon already on hand? These questions must address themselves to our notice in every step of our progress. Previous chapters have so fully established the fact that there was an annular system, which in part remained on high till man came upon earth, that all we now need to do will be to show more clearly that that system was filled, as it were, with unconsumed carbon.

I suppose there is no man of reason who will upon mature reflection deny that the earth was once in a fiery molten condition. But if the earth ever was in a molten condition, can we possibly avoid the conclusion that it was a smoking world? Suppose this earth, or any orb constituted as this is, should be by some means suddenly changed into an igneous or glowing sphere. That the immensity of carbon contained in its rocky frame would be under such conditions driven outward in the form of smoke, or unconsumed carbon, will not be questioned by the philosophic student. And is it not an axiomatic fact that that ocean of expelled carbon would possess all allotropic forms of that element from the heaviest to the lightest? Is it not an axiomatic fact that these forms of carbon would, to a great extent, arrange themselves in that ocean of vapors in positions determined by their specific gravities the heaviest nearest the earth? My readers certainly would not require me to supply witnesses to prove these things,—questions so evidently true that no further evidence can add to their force.

Hence, it is a conclusion to which we are all compelled to assent that the primitive atmosphere was largely an ocean of distilled carbon; and as we know that nearly all unconsumed carbon arising from every furnace is a fuel (since it unites with oxygen and disappears), we know that that primitive carbon in some of its forms was also a fuel. And as we know (as any one can prove by experiment) that carbon particles or atoms in their nascent state rising among aqueous vapors will decompose them and unite with their oxygen forming an invisible gas, and also with their hydrogen forming a hydro-carbon,* then we also know that the primitive atmosphere contained a fund of fuel hydro-carbon. Soot is deposited in infinitesimal smoke particles. Hence we know that smoke from burning carbon is simply a fuel. And we are thus, by the inexorable demands of law, driven to the conclusion that the smoke that arose from the igneous earth was a fuel hydro-carbon; and further, that when the primitive vapors, segregated and aggregated into an annular system, these forms of carbon were present in that system, and also that when

* Even soot, placed away in a vessel of water, will in time decompose the latter, appropriating its oxygen and a small part of its hydrogen.

the same rings declined from the annular form into terrestrial belts, this same carbon was present, overcanopying the earth on its way to its surface, near the poles, or at least beyond the temperate zones.

Now when we turn our eyes to Jupiter and Saturn, and see their dark belts in perpetual motion *inter se*, and can find nothing in the whole laboratory of nature competent to produce such belts, except carbon, and know from analogy that these planets also have been burning and smoking worlds, we simply see the Jovian and Saturnian carboniferous strata revolving as annular matter; and the process of primitive carbon distillation becomes a universal one in the economy of world-making.

There is yet another feature to be examined before we are quite ready to examine the coal beds. I presume that every one of my readers can see that if the condition of the primitive or annular carbon be true, as here predicated, the annular theory is being narrowed down to a few decisive tests. It must be seen that if the annular matter arranged itself in the system according to its specific gravity, then the heaviest forms of carbon such as graphite were located nearest the earth, and therefore fell to it long before the lighter forms. Now, this being the case, where must we find this heavy carbon? Certainly in the first-formed aqueous beds. That is, if the annular theory be true there must be found in the archæan beds vast quantities of carbon of the greatest specific gravity. Now this carbon having fallen directly after the igneous era closed, must be found unassociated with fossil vegetation. And if thus found, it becomes absolute proof that carbon, contrary to the opinion of geologists, was formed without the aid of vegetation, and consequently supplied directly by the primitive process in the igneous earth. Now let us hunt this primitive carbon to its hiding place. If we cannot find it, the annular theory is a failure. If we do find it, it is once more triumphant; and if found unattended by a fossilized vegetation, the thēory will again be vindicated.

At this stage of the argument, then, it is with delight that I turn to the highest of human authority, and find the wished-for carbon away down amid the primitive piles of aqueous beds. Dana tells us graphite, a form of carbon, is "very common material"* in the oldest beds. Prof. Dawson claims that "the quantity of carbon in the laurentian [oldest beds] is equal to that in similar areas of the carboniferous system."† It is mined for graphite in many parts of both hemispheres, and sometimes occurs in massive beds; sometimes forming from 20 to 30 per cent. of the laurentian lime-beds. It is an important constituent of some of the iron strata of the archæan of the old world. Some of the purest and best deposits are found in the archæan of New England. That is, enough is known of these oldest beds to establish beyond a doubt the claim that they contain the very material we are in search of to vindicate our theory. And as we have previously shown that these heavy mineral and metallic beds were largely an annular product the presence of heavy carbon in some only adds strength and philosophic value to the claim.

But where are the evidences of vegetable life? Where can we find a fossil bud, leaf or stem? For

---

* "Manual," page 152.

† Ibid., page 157.

fifty years an army of geologists with eyes open and keen as the eagle's, has explored the archæan world in vain for a satisfactory trace of a vegetable. Can it be possible that plants grew on the old world shores without leaving a trace? Why did the delicate forms of the eozoön leave their impress on the imperishable granite without a trace of a plant? Simply because no plants were there! Then how are we to avoid the conclusion that amid the oldest sedimentary beds are vast deposits of carbon in whose formation the plant never took a part? Dana says emphatically "no distinct remains of plants have been observed" in these ancient beds; and as the carbon beds themselves are the only evidence found, and these no evidences at all, where is the hope of the vegetarian?

All geologists admit that if coal be a vegetable product graphite must also have had a vegetable origin; compromising only so far as to admit that animal organisms may have aided in the process, which of course only adds to the difficulty, since it is carbon that makes the organism, not the organism the carbon. Here, then, is a problem which the vegetarian can neither circumvent nor climb over without the aid of the annular theory. The foundation stone upon which the vegetation theory stands has vanished in primitive fire, and the whole edifice tumbles into a mighty mass of ruins. Here we are compelled to admit that the graphite is a primitive carbon; that carbon did exist, and was placed as a sedimentary bed in the earth before a plant ever grew upon its surface. Hence the plant did not form the carbon, but the carbon formed the plant. Upon this eternal plan the world was built. From the carbon beds locked amid the metallic and granite sills of the earth's crust to the peat swamp of the present day, carbon has been king, and the plant its pliant product.

But suppose an abundant vegetation did exist in archæan times, could it in the least invalidate the philosophic claim that graphite found in the laurentian beds was derived from the great telluric furnace? Inexorable law demands that graphite carbon must be found in the oldest sedimentary rocks. There it is. It also demands that it must there be found as an igneous product. The entire absence of organic fossils asserts that it is an igneous product. Now suppose in the coming centuries, some leaves, some stems or other forms of vegetation should be found in graphite; these would, according to law, become graphitic, merely because they were imbedded in graphite, for the same reason that if the plant form had fallen in a sand bed it would have been a silicious fossil. This law must hold good at all times. Now if men should find abundant vegetable fossils in graphite it is simply ridiculous to argue from this fact alone that graphite is of vegetable origin. There is an abundance of vegetable fossils in clay beds, and in sand beds, etc., but who would claim from this fact alone that the clay bed or the sand bed in which they are found is of vegetable origin? The simple fact that organic fossils are found in carbon beds, and changed to carbon, affords no evidence at all that those organisms made the bed. They are simply carbon fossils because they were imbedded in a carbon stratum, for the same reason that fossils found in a lime bed are calcareous fossils. Human remains have been found in calcareous formation, but it does not follow that limestone is of human origin. And yet in defiance of this very law, regulating the fossilization of organic forms, men claim that the carbon strata of all ages are of vegetable origin because vegetable fossils are found in some of them. Thus it must be admitted that he who claims that the graphite of the archæan strata is of vegetable origin advocates the miraculous suspension of natural law; first, because there was no vegetation existing at that time,* and second, because the law requires the existence of carbon in those strata that is not of vegetable origin.

The reader can now judge for himself which theory is supported by the facts. He can also see that the vegetation theory is mortally weakened by the simple fact that the earliest carbon beds cannot by any possibility be of vegetable origin.

Men may call the non-existence of vegetation a negative evidence; but since its existence is no evidence, either positive or negative, that the bed is a vegetable product, of what value is it to vegetarians? Even as we enter upon paleozoic time we look in vain for any forms of vegetation, but the very lowest cryptogamic species, and these in very scanty exhibits, and also marine in habit.

Now as graphite is not a vegetable product, it is very probable that other forms of carbon, as bituminous and anthracite coals, are not. Considering that we must reverse the law of fossilization in order to conceive of any stratum itself made out of the fossils it contains, it is scarcely possible that a coal stratum can be a vegetable product.

* Dana only echoes the universal opinion of geologists when he says: "No distinct remains of plants have been observed" in the archæan rocks, and as even in the huronian no satisfactory traces of plants have been found, we are safe in the claim that there are no plant remains in or among the graphite beds.

We find vegetable remains in coal seams just as we find them in any other rock. Sometimes a coal-plant, as a lepidodendron, planted in the under-clay rises through the coal bed and extends into the overlying shale and sandstone. But here we find it a clay fossil in the under-clay, a carbonaceous fossil in the coal bed, a silicious fossil in the sandstone; that is, if it has at all become mineralized. Now the very presence of an upright stem, or a trunk of a tree, in such beds is proof positive of the rapid accumulation of the beds around it. A tree standing while 5 feet of vegetable carbon accumulated around it indicates a fall and accumulation of 40 feet of vegetable debris. Can it be possible that a tree would continue to grow in a swamp or marsh while the growth of vegetable matter sufficient to make a bed of 40 feet in thickness is deposited at its base, and then continue to stand till massive beds of sand and clay are deposited upon the layer of carbon, according to the usual slow process of strata-building? On the supposition that such accumulations are deposited, as they now are, we are forced to face the miraculous. On the supposition that the carbon fell from the annular system we are led to the conclusion that a bed of carbon five feet in thickness might accumulate in a few months; nay, it might be in a few days.

There is one important feature that has been greatly misapprehended by geologists in considering the coal question. It is a fact easily demonstrated that the vegetable carbon in the coal beds is generally not bituminous even in a bituminous bed. We often find a thin layer of vegetable carbon in the solid coal. It is an accumulation of vegetable debris carbonized. Any one who will take the pains to collect this vegetable matter can readily satisfy himself that it is scarcely combustible. If a plant should fall in a bed of carbon, and afterward, by the aid of pressure and heat, become saturated with oil, or bitumen, it would thus be made combustible; but nine times out of ten the true vegetable matter found imbedded in coal burns with difficulty. Such is also the case with true lignites and vegetable peat. Now this could not be the case if the coal beds were made of vegetable carbon; for the abundance of bitumen in the oily coals necessitates that the vegetation should contain the elements of the same. And if the coal plants contained a resinous sap, as is now claimed by some scientists, even the vegetable charcoal would be bituminous. Thus the very fossil vegetation speaks plainly in opposition to the vegetable origin of coal. Now why is it the very plants which geologists claim are necessary for the formation of a bed of coal, when gathered from the body of the coal, will scarcely burn if these plants formed both the body of the coal and the bitumen or oily matter which exists as an essential part of it? The conclusion, it seems to me, is inevitable that the vegetation found in coal is to a great extent foreign matter, just as the ferns so abundant in the clays over the coal are foreign to the clay-beds,—i. e., simply an involved vegetation. The fact that coal has been considered a vegetable product, and the statement that vegetation is found in coal are so misleading that the common reader has the impression that plants and the remains of plants are found in abundance in a coal seam, while the fact is that in many coal veins there is a paucity of vegetable matter observable by the naked eye, and in some coal veins it is almost entirely absent. For one visible plant impression in the coal itself there are ten in the roof of the coal. Did this great abundance of vegetation give rise to the clay beds and shales composing the roof? They are clay fossils because they were imbedded in clay, as the fossils in coal are carbon because they were imbedded in carbon. And the simple fact that there are fewer fossil remains in some coal than in the superimposed beds, shows that the carbon occupied less time in collecting and forming into a bed.

Again, it is claimed that the original form of coal is that of ordinary peat, now accumulating in peat bogs and marshes. I have given no little attention to the formation and nature of peat beds in a former chapter, but I wish to say in this connection that if peat were compressed as coal has been the delicate lamination so prevalent in coal would not exist. I have frequently counted a dozen or more leaves or laminæ in one vertical inch of coal; sometimes these are not thicker than brown paper. And in the examination microscopically of a vertical section of coal no fibers can be seen running through the mass, as would be the case if coal were compressed peat.

A mass of peat has been compressed with a force of 20 tons to the square inch, and yet the vertical structure of the mass was apparent. Now it matters not, I presume, what amount of pressure is employed; it cannot make the fibers, roots, carbonized twigs, leaves and stems that are well known to "run up and down" in a mass of peat to change their position, and lie horizontally as they do in a coal bed. And this would be the more evident as the mass of coal was the more extensive. Now lest some of my readers think I press this view too strongly, I will call their attention to Dana's description of peat,* who says it is "commonly penetrated by rootlets." Will some one tell us what amount of pressure is required to make all these rootlets lie horizontally in the coal? It is very rarely that we find any roots and rootlets in any position in the coal itself, and they are much more rarely found running vertically or across the laminations.

Since we find an abundance of rootlets in the underclays of coal, running in all directions, vertically as well as horizontally, it seems conclusive that coal is not metamorphosed peat. Now imagine a world filled with marshes and peat beds; not like the thousands of peat bogs that are found on every continent, but great continental coal marshes 10,000 or 100,000 square miles in extent. Imagine these marshes but little or any above the sea level, and covered with calamites, ferns, sigillaria and lepidodendra,—plants of the carboniferous era,—and after remaining for countless centuries as a motionless continent, to suddenly sink beneath the waves of the sea, in order to receive a sea-formed bed for a covering; and in this universal burial to preserve but a paucity of vegetable fossils, and these mostly in horizontal laminations,—while in the clays immediately under as well in those immediately above the coal to be a profusion of fossilized vegetation. This is the character of some of the coal formations.

Now in order that a second coal seam should be formed after 20, 50 or a 100 feet of clay, sand and lime has accumulated over the buried carbon bed, this great expanse must arise just as high above the waves as it stood for the first vegetation. If the accumulated sand was 20 feet, then the first formed coal beds arose to

* "Manual," page 616.

within 20 feet of the surface of the sea. If 50 feet of sand, clays, etc., had accumulated over it, then it arose to within 50 feet and then ceased its upward motion, and remained a fixed expanse for ages, until another bed of coal had accumulated, and then sank again. Now if these changes had taken place but once or twice we might conclude that as it was a remarkable coincidence, chance might explain it. But when we are forced to add miracle to miracle by admitting that these changes took place to a large extent simultaneously in all lands, in all continents, our credulity becomes unduly stretched. And when we are compelled to admit that this oscillation of sea and land lost its regularity at times on some continents and was repeated from fifty to one hundred times to form the coal beds of the same, and that the submergence was frequently to abyssal depths in order for the accumulation of limestone strata, the question amounts to a ridiculous absurdity.

It is plain that if the oscillating bed had arisen as far above the waves as it sank again and again beneath them that no swamp vegetation could have accumulated. But how did it occur that it stopped so often just in the right place, and became so often a permanent fixity in an age of constant unstableness? How did the beds of other lands join simultaneously in this process?

Thus it seems that the coal is planted immediately upon an aqueous-formed stratum, and all the beds between coal beds are also aqueous strata, and as we must find some primitive carbon beds as aqueous formations here in the midst of sedimentary beds we will put in our claims. Beneath the peat-bog carbon of this age there is frequently a shell bed of fresh-water species, and the carbon is deposited immediately upon it, while the fire clays and dirt beds of the coal formations show that the conditions of carbon accumulation were entirely different at the very beginning of coal-forming. Peat forms in fresh-water swamps. But the coal strata, locked between marine beds, show that if they are a vegetation the plants were marine. But right here we will quote from high authority.* "Algæ can therefore produce nothing in the shape of coal." . . . "They cannot burn or emit any amount of caloric." . . . "*All remains of plants found either in the shales which cover the coal beds, or in the body of the coal itself are land plants, . . . none of it is of marine origin.*" (Italics mine.) Now what are we to do? The same authority states: "The remains of the plants are only found in the roof-shales of a coal bed."† From this we are forced to conclude that the vegetation thus preserved as fossil is simply an involved one, and came into the roof-shales of the coal after the coal-bed was submerged in the sea, and we simply have no means at all to show that the coal is a vegetable product. Thus every step we take leads us deeper into difficulty. We must find some way to account for the fact that coal was deposited over a vast area of sea bottom in different regions at the same time. We must find a process that abnegnates the ridiculous and accommodating submergence and re-elevation of beds a hundred times repeated, which the old theory necessarily maintains. We must find a process of fuel-formation in beds that contain but little visible vegetable remains, locked, as all marine formations are, between marine beds. Thus, as we take a comprehensive glance at the difficulties,

* "Leo Lesquereux Report of Prog. Pa., page 609.

† Ibid., page 618.

we find the vegetation theory to be unnatural, and not to be admitted, whether any other explanation is within our grasp or not. "A half-way explanation will not do." And an unphilosophic explanation cannot be tolerated.

Now I wish the reader to understand that I do not oppose the idea of local submergences, for such things do naturally occur. But I do claim that the whole world could not have been flooded except by a down-rush of super-aerial waters. The structure of the continents proves that when once formed they forever continned to be continents.

Dana, in referring to the grand structure lines and frame-work of the continents, is forced to say: "There is strong reason for concluding that the continents have always been continents; and that while portions may have at times been submerged some thousands of feet the continents have never changed places with the ocean." Now if there be any truth to be derived from the carboniferous conglomerate beds, it is the fact that they were synchronously formed the world over. It would be as vain to deny this as to assert that the deposits of the last glacial period were not formed all over the Northern Hemisphere during one and the same period, extending through unknown centuries. But the intimate relation of these conglomerates to the coal-beds shows that both were involved in whatever submergence or whatever change of level took place.

It must be apparent that coal veins formed, according to the general view, in swamps of vast extent, must have a general parallelism; and this is the view generally held by geologists. On the other hand, if coal be an aqueous deposit upon the sea bottom it is plain that the parallelism depends entirely upon the thickness of the intercalated beds in different places. Suppose a downfall of carbon dust should occur to-day. Borne away into the ocean it would settle upon its bottom over all its irregularities and its plains, of course subserviently to directing currents. Then the sand-beds accumulating for ages are placed upon it. These, of course, would form a greater thickness in some places than in others. Hence a succeeding fall of carbon settling upon the ocean's floor could not form a bed parallel with the first. It is a matter of ocular demonstration that there are actually no such things as parallel coal veins. Sometimes for short distances they appear to be so. In my own neighborhood the distance between the several coal seams varies from 20 to 40 feet in less than one mile. The main coal seam of the Leatherwood Valley, five miles west of the Barnesville coal shaft, is ninety feet lower at the former place than in the shaft. As there is no strata fracture here these beds are evidently lying now as they were placed, which at once refers them to aqueous formations on the sea bottom. A careful measurement of hundreds of localities, given in the geological surveys of the different States and Territories, as well as of Europe, demonstrates it beyond a peradventure that there is a general and universal want of parallelism among coal-veins. Prof. Newberry has shown this so clearly in his report on the Ohio coals* as to leave no room for doubt that the coal beds did not thus accumulate. Hence, we are again driven to the only other source—i. e., an accumulation of carbon upon the undulatory floor of the sea.†

* Volume II, "Ohio Reports," pages 126 and 169.

† While I am presenting this feature, I must, even at the risk

Cannel coal is another unimpeachable witness of the aqueous deposition of carbon beds. It is admitted, I believe, by the principal geologists of America, that cannel coal is derived from vegetable matter completely macerated in water, and therefore actual aqueous sediments.

Within three miles of my dwelling is a mine of coal, on one side of which the formation is bituminous, while not 100 yards distant the coal is true cannel, and the gradation from the bituminous to the cannel region is so gradual that it is impossible to tell where one terminates and the other begins. The annular theory would make both an aqueous sediment, and just as the fine sand would separate from coarser by the direction of currents, and form a bed of its own, so would the finer particles of floating carbon and clay separate and form a bed of cannel coal by the side of a bituminous bed in such a way that no man could point out the line of transition.

I am glad thus to be able to quote so reliable authority on this subject, so vital and yet so fatal to the vegetable theory. Prof. Andrews, with the keen eye of a practical philosopher, saw that the coal-veins must be parallel, or the vegetation and submergence theory was opposed by law. At least he knew very well that all

---

of prolonging this discussion, give Prof. Newberry's opinion on the parallelism of coal seams. He says: "Prof. Andrews accounts for this claimed parallelism by supposing that the different coal seams were formed at or near the line of water level, and that the subsidences which have caused the successive layers of carbonaceous matter were continental or uniform. To these views I have been unable to subscribe, inasmuch as I have failed to detect the parallelism claimed, and on the contrary, have, as it seems to me, in numerous instances, discovered very marked inequality in the distances, that at different localities separate coal seams which are unmistakably continuous."

continents could not simultaneously plunge into the sea again and again, and not maintain a general parallelism. Prof. Newberry, with the rich stores of an indefatigable and correct observer, announces the very opposite conclusion, and backs his views with an array of facts that none will dare dispute. I cannot follow him in detail, but must refer the reader to the report itself. But when he tells us that coal, No. 1, is so exceedingly variable as to show a series of waves, whose summits are in many places 50 feet higher than the trough "within the limits of a few hundred acres"; again, when he states that in the northwest corner of Carroll County, Ohio, within an area of 1,200 feet, the distance between coals, No. 3 and 4, varies from 20 to 45 feet, and at another point from 20 to 90 feet, and another 110 feet, and when he says that the interval between No. 4 and No. 6 is equally variable; that the intervals between No. 6 and No. 7 vary from 54 to 100 feet; that the distance between the great Pittsburgh seam and the Ames limestone varies from 140 to 225 feet; and again when he states "it has been proved that between Barnesville and Bellaire the space between coals No. 8 and No. 10 increases by more than 100 feet," he only states a philosophic and necessary fact.

We will consider the famous cannel mine at Cannelton, Pennsylvania, as a representative of this class of coal. In the bottom of the mine is an 18-inch bed of pure bituminous coal. On the top of this is a heavy mass of true cannel. Here, then, as usual, we are at once confronted by the unnatural fact that two veins of coal, one placed immediately upon the other, without one-tenth of an inch of vertical gradation; not the space of the thickness of common writing paper between them, so as actually to make one solid seam of coal, yet one part is claimed by high authority to be an aqueous deposit and the other not. The bituminous is made to be a vegetable bog-growth, and yet is clear of any permeating rootlets; shows hundreds of fine laminations, just such as one would expect to see in a fine sedimentary deposit from water. The cannel, a massive accumulation of black carbonaceous mud, with very little evidence of lamination, is made to be a completely macerated mass of vegetation. The bog, then, in which the bituminous mass was accumulating, sank not as slowly as is the natural process of to-day, but suddenly from the swamp level to the condition of a lake or pond, and the hypothetic slow and long maceration began instantly to supply the cannel carbon. There was simply no time given for "maceration" before the vegetation was macerated, and began to fall on the bituminous bed. No conscientious geologist can stand in this cannel mine, seeing these things as plainly as the light of day, and say, this is the tale of the cannel! The bituminous bog could never have sunk and received the cannel vegetation as thus claimed, and when I find this the case in hundreds of places, sometimes the cannel above, sometimes beneath the bituminous and so often is it the case that no parting exists between them, that we may call it the rule rather than the exception. The subsidence, if it ever took place in these cannel beds, should have made a parting of something that in-rushing waters must have conveyed to the spot. Thus, while it is impossible to explain this sudden change in the character of the two deposits by the old theory, if we will but admit that the same process that plants a bed of limestone immediately upon a sand bed without any signs of gradation—by simply depositing the bituminous carbon first, and then by a slight change of moving currents bringing in another form of carbon, the solution is plain.

We must conceive this carbon as susceptible of transportation and change as any other sediment, and local beds of fine carbonaceous mud, which the cannel carbon really is, could not avoid formation while currents ran, any more than sand or clay. Admit all such beds to be sedimentary deposits of annular carbon and every mystery in their formation vanishes.

Black carbonaceous shales, so universally prevalent, must have had this same origin. Sand and clays mixed by intermingling currents with floating carbon in black carbonaceous waters could not fail to give rise to black shales, and as many of these shales are almost devoid of fossil organisms, and especially of plants, to attribute their color and the presence of carbon to vegetation is unnatural. And all must see that if the earth was ever in a molten state, its annular carbon, falling in after ages as so much primitive soot, could not have failed to blacken the waters of the ocean wherever currents moved, and the sediment deposited therein must have mingled with the carbon. If the carbon had been in excess in these shales the result would have been a deposit of cannel coal; on the other hand, if the clays had been in excess in the Cannelton mine, the result would have been a black shaly deposit. If no mud had interfered, the whole deposit would have been a bituminous coal.

A remarkable deposit of coal exists in eastern Pennsylvania, at Summit Hill,—a spot made historic by the great Lyell. Here seven coal veins at first occupying a vertical range of 134 feet (including intervening strata of rock) so rapidly approach each other that they all unite into one seam in less than five miles from their beginning. Now, according to the old theory, that part of the bog where the one heavy seam is, stood firm and was never once submerged, while the space immediately adjoining went down and returned seven different times until 134 feet of rock were intercalated and seven seams of coal were formed out of seven successive swamps, and finally the whole swamp went down together. Every man must see how unphilosophic this is! And yet how easily explained as sedimentary beds. A carbon bed was formed, currents carried other matter and covered up a part of the bed. Another supply of carbon settles down upon the whole, and the process is repeated again and again; the strata having thus been formed as other strata are at this day. This simply is a refutation of the swamp theory, and the idea of the parallelism of coal veins.

It is well known that "quite a number of boulders of rock foreign to the localities where found have been met with in the coal seams of Ohio." (Newberry.) Here is another emphatic and decisive test. Boulders in a coal seam mean a coal seam formed under water, and a foreign boulder in a coal seam means a coal seam formed at the bottom of the ocean.

The vegetarians so far concede this as to admit that "the ocean must have been very near," that "the ocean must have made an inroad upon the coal swamp," etc. Now since the ocean must have been "very near" to have deposited the bed upon which the coal seam was placed, and "very near" to have formed the bed placed immediately upon the coal, and very near, when boulders were floated over its surface and dropped in the coal deposit, let us admit the inevitable truth that the ocean was "very near" all the time!

And since the ocean was so "very near" as to permit some floating body, as a tree or moving ice, to drop a boulder into a forming coal bed at the bottom of the sea, it was about as close as we need to have it in order to crush the swamp theory forever.

In conversation with the intelligent proprietor of the Cannelton coal mine, above refered to (I. Mansfield), I learned that water-worn pebbles had been found in the coal there. A boulder now in the museum at Columbus, Ohio, and found in the middle of a coal seam at Shawnee, "weighs not less than 200 pounds," and showed the marks of glaciation. The coal above this boulder "was normal in all respects,"* showing that the vein finished forming after the boulder fell into it. Now it is plain that if there had been a submergence and a re-elevation of a coal swamp that the condition of the vein would have left indisputable evidence of the change, but as all the evidence is positive and directly opposed to a submergence, and in favor of a continued and uninterrupted deposition of carbon upon the boulder and its new surroundings, it is impossible to avoid the conclusion that the Shawnee coal seam was a sedimentary aqueous formation,—one of the seams that the igneous world under the dictum of law declares must be found in the aqueous crust. Boulders foreign to the locality are found at Carbondale and Nelsonville, and I have been repeatedly informed by intelligent miners, who have worked in Scotland and on the continent of Europe, as well as among the Rocky Moun-

* "Ohio Report for 1884, Vol. V, pages 136, 1006.

tain coals, that they had met in coal mines with water-worn boulders and pebbles (not concretions) of different colors. When we take into the account the numbers that never came before the gaze of the geologist, and remember that but a small part of the field has ever met the eyes of man, we may be allowed to magnify the evidence these boulders afford. If men are not bound by the merciless fetters of antiquated opinion they must see that the vegetable theory has here an obstacle that they cannot remove.

But other evidences are found in the coal veins quite as positive as that of the boulder. There is scarcely a coal vein in the United States, to say nothing of other countries, that is not divided at least once, and many of them are divided into many horizontal sections by clay seams that were deposited from the ocean's waters while the coal seam was forming. These clay partings are generally very persistent, varying in thickness from a few inches to that of writing paper. In some seams these extend over thousands of square miles, and few of them contain vegetable fossils; a few of them animal organisms; but the greater part are wide reaches of barren clay. These seams declare what no man will contradict, that all the coal veins of the United States, during some period at least, and many of them during frequent periods, were under the waters of the sea. I only go one step farther than my brother geologists, and instead of admitting that "these things demand that the oceans were near at hand" when these clays were deposited, I admit that they were present, not only when the clays were deposited, but also when the coal veins containing them were deposited.

But the testimony of these clay-partings does not stop here. Supposing that the lower bench of a coal seam really was a vegetable product, "grown in situ," with its root bed below the coal seam, as is claimed; after a seam of carbon is thus formed the swamp is submerged, and an outspread of mud covers up the vegetation to the depth of half an inch without enclosing a trace of that vegetation in the parting itself. Now it is simply impossible that a parting of clay should settle down upon an expanse of submerged vegetation without preserving that vegetation in itself. But it is the rarest thing that even a trace of a plant, stem or leaf, is observable; what conclusion, then, must we draw?

But how in the name of reason did the vegetation that formed the next bench of coal take root in this thin seam of clay? Have the roots of this succeeding vegetation been found in this thin clay-parting? I have never known of such an instance, and yet I have lived among coal mines the greater part of my life, and have carefully examined hundreds of localities for them. We are plainly forbidden by the evidence and the verdict of law to claim either a submergence or a vegetation. But suppose that to-day a great carbon fund should float from the Arctic Ocean into Hudson Bay. This carbon would settle upon an undulating bottom, and if a flood of muddy waters from the surrounding rivers should empty into the Bay while the carbon bed was forming, a thin clay parting over wide areas would inevitably follow. The clay being heavier than the carbon would immediately settle, and allow the carbon to complete its deposition afterwards. And further, as this inpouring of carbon from the ocean might continue for years there would be ample opportunity for many clay seams.

Now it must be admitted that if there ever was carbon dust in the annular system; in short, if there ever was an igneous and smoking earth, such carbon veins with such clay-partings do now exist in the aqueous crust. Have we not found them in the United States coal-formations?

Microscopic organisms in coal so far from being opposed to the theory here advanced become valuable aids to it when intelligently considered. I have elsewhere referred to the axiomatic claim that the annular system contained the seed beds of animal and vegetable organisms. It is beyond our power to determine how far the evolution of organisms continued in this annular world. Trusting that others more capable may in the future come to a philosophic conclusion in the matter, I will not press it further than to reiterate the necessity of the claim that the floating mass of primitive carbon clouds, after they entered the atmosphere and floated away, perhaps for centuries, toward the polar regions, in their efforts to reach the earth, became a tissue or web of evolving vegetable organisms, accompanied with an immensity of microscopic forms. Now I know not to what extent microscopic forms exist in the mass of the coal. I only know they should be there, and that in the deposition of the coal they should be segregated upon definite surfaces. Let it be understood that I am not drawn to this conclusion from any suspicion that criticism will force me to it, but from careful study and long research in experimental work. I have taken fresh soot from the furnace, within a few minutes after it was formed, subjected it to the hot vapors from boiling water, and stored it away in an open vessel of water, and have seen vegetable and animal organisms start into being, live, propagate and die therein. This same experiment any man can perform. Now this having been done, I want to know what is there that can possibly hinder floating or revolving soot-clouds in attenuated air, or even in the annular system, —an ocean of matter where every disposition and poteney of matter existed,—from being regions of organic development. For this reason it is impossible for me to look upon the belt system of Jupiter as any other than an ocean of organisms, adapted to their own peculiar environments. It seems to be the inevitable and universal summation of a disposition in matter, akin to spirit, under a controlling intelligence. Then looking back at the carbonaceous downfalls, ranging through countless myriads of years, I behold floods of microscopic and other organisms, I know then, experimentally, that carbonaceous waters are favorable to the evolution of organic matter, and theoretically that carbonaceous clouds are also. What other conclusion, then, can we come to than that a fund of carbon floating in the ocean, or lying in lakes and ponds, would give rise to vegetable and animal organisms adapted to the conditions existing; and that this carbon as it reached its destination, in the tedious and protracted round and circulation of currents, would involve these forms in the accumulating mass?

If the clay mud at the bottom of lakes, or the calcareous ooze on the ocean's floor, as it accumulates into beds, involves its own peculiar life forms, and presents a mass of microscopic and other organisms, I cannot for any reason see why the carbonaceous ooze in the same lake beds, on the same sea bottom, could fail to involve its characteristic forms, and that the same as fuel carbon would not to a large extent exhibit them. Hence, I predicate that future researches in this direction will reveal the fact that the mass of coal contains just such organisms as the ancient carbonaceous environment demanded.

As is well known, an abundance of marine vegetation exists upon the sea bottom in all favoring localities—creeping stems, with roots and leaf-like forms; floating vines with reaching tendrils, and with roots fixed in the mud. A carbon sediment rapidly accumulating would involve all this.

Now the geologist will not fail to see that another important question is here involved. Under almost all the carbon veins there lies a bed of fire clay,—a "dirt bed." It is a little strange that immediately adjoining a highly combustible bed a substance should be so invariably planted so refractory as to form crucibles for the fusing of almost every known metal. In this bed lies entombed a profuse marine vegetation, and the fact that its delicate lineaments have been so well preserved proves that it was suddenly involved. The fact that it is practically infusible argues that it was a fire-born distillation of primitive times. The fact that it so generally accompanies a primitive carbon product argues the same. The fact that it is more generally present under coal veins that are more distant from the tropics, and invariably present in the most distant ones, leads to the same conclusion. While the fact that fire-formed clay dust, sublimed in the great telluric crucible, must have arisen and commingled with the primitive vapors, and returned with them, impels to the conclusion that when a carbon-fall occurred, this clay matter, necessitated by its greater specific gravity, separated and fell first upon the ocean's floor.* This fire clay is found in a modified form under beds of primitive graphite where no vegetation is involved, and therefore cannot be a vegetable distillation. It is found between massive beds of glacier polar ice, immediately under a carbonite deposit, as at Kotzebue Sound, which enforces the same conclusion. This fire-born clay is found in such stupendous masses in almost all glaciated districts, whose glaciers radiated from polar regions, as to utterly confound the geologist in his efforts to find a philosophic source. It is found in lands a thousand miles from abraded mountains; in beds which prove by analysis to be peculiar, and not what we would expect, as mud pulverized by the moving ice. The persistency of these peculiarities in all lands, whether the abraded region be silicious or calcareous, micaceous or feldspathic; whether the neighboring hills and mountains could yield such clays or not, must, it seems to me, lead us to look to the annular system as its source. It is said that not one of the more than 70 coal beds in the Nova Scotia region is without its characteristic clay bed. When we see trees standing in and surrounded by this clay, and rising through the coal seam, and even penetrating many feet into the overlying rock, we are forced to admit a rapid accumulation. So that every standing tree in such a position, so far from evincing the claims of vegetable distillation either in clay or coal, during immense periods of time, stands as insuperable evidence against them. These beds accumulated during the

* I greatly regret that space does not permit me to follow this important question further. A volume might be written on these fire-born clays alone.

lifetime of the tree, and not only the coal, but the sand beds above, accumulated around it; and reason urges the claim that the tree could not long survive this ordeal. But the most moderate estimate of the time required for the slow vegetable accumulation of a coal bed alone, it seems to me, ought to settle this question. On the other hand, admitting that every annular downfall of matter, aqueous vapor, carbon-dust or snows, must have brought their associated clays, and more largely in the more recent falls, all these otherwise intractable problems are explained. The tree growing in its own swamp clay bed, first involved by the augmented ocean, must have perished then. Immediately the ocean, muddied by the clay and carbon, precipitates first the clay as the normal bed for the coal; then follows the carbon; and the terrific flood of waters from adjacent continents brings in the muddy waters, and the great fund of vegetation of the upper clays; then upheaval, and other changes resulting from oceanic augmentation, supplies the other assorted beds. This brings us to that point when we must consider the question of lime and sand strata and their fossils, as referred to in a former chapter.

I know not whether standing trees rising from coal veins have ever been found in regions where limestones predominate. The annular theory requires that such should not be the case. These strata are largely a deep sea formation, and only such as were deposited as mechanical precipitates could be formed in shallow waters, especially in regions beyond the tropics. Hence it is plain that if we find an abundant fossil vegetation in the clay beds above the coal in limestone regions their presence antagonizes the new theory, and if not found it antagonizes the old one. While also a limestone stratum deposited, manifestly near and among shore deposits, or continental detritus, points directly to an annular origin, and here we will expect to find vegetable fossils in the upper clays. Now geologists have here a chance to prove or disprove this problem. So far as my observations have extended in the Appalachian coal field this theory is abundantly sustained by the great limestone strata. And, again, so far as I am able to gather evidence from the surveys of the western coal fields, it is felicitously supported by the general absence of lime beds among the coal seams, and the presence of immense amount of vegetation as fossils in the interposed beds of sand and clay. Again I must abruptly close the consideration of this wonderfully rich field of thought.

I have several times referred to these as important tests of the truth of the annular hypothesis. Since it is self-evident that peat vegetation for the distillation of carbon fuel could never have assumed a foothold in any region if the peat foundation had not been previously laid down; and since the great peat-forming regions of the earth increase in importance and extent from the region of the tropics toward the poles until we find them under the polar circles where the soil is solidly frozen the year round (except a slight covering of soil in the short summer), it is plain that some stupendous supply of carbonaceous matter has been added to these colder regions in modern times. I have shown why annular matter must fall in these higher latitudes, both north and south. I have shown that carbon must have constituted a part of that matter, and how that the more recent downfall supplied it, as an accompaniment of snows, etc., but it seems necessary at this point to give some further facts tending to corroborate the claim that the peat foundation is what I have called carbonite as distinct from peat itself,—an annular product.

In Dr. Anderson's "Practical Treatise on Peat-Moss" he frequently alludes to beds of "black sediment," "rich loam," "black moss" sunk beneath the waters of the sea, lakes and ponds. Sometimes divers have reported "black mud" in the Scottish seas from 100 to 180 feet deep. At Loch Alsh, also at Oban, more than 100 feet beneath the surface of the water, the harbor bottom is covered with the same black carbonaceous matter. The same is seen at low tide on the shores of the Frith of Forth.*

What is more, this same black carbon deposit "has been dredged far out in the German Ocean."† In nearly all these carbonite deposits, whether on high ground, or in the sea, are found trees prostrate, and some in the seas with the trunks erect. And the simple fact that some of these trees are such as do not grow in peat bogs, is the strongest kind of evidence that they are not true peat bogs. The ash, oak, fir, alder, etc., etc., says Geikie, are found in them, "rooted in the kind of soil they are known to prefer!" ‡ These trees are sometimes so well preserved that they are made into merchantable lumber. And as peat is an exceeding slow formation, how did they become involved in peat beds and remain undecayed, even if they could have flourished in a bog? This carbonite, so-called peat, is

* Geikie's "Great Ice Age," page 298.

† Ibid., page 300.

‡ Ibid., page 294.

not confined to the lowlands. Large areas of Scottish Highlands are covered with it, just as vast portions of Russia, Siberia, British America, and the United States are, and which if covered by earth would be peat to all intents and purposes.

But who has ever reported peat submerged in the tropics—those very regions where luxuriant vegetation grows and dies? Is there any philosophic reason why peat should not be found more abundantly there than elsewhere, except the fact that the foundation carbon has never fallen there? And where in the tropics can the geologist point to any important coal beds? Here is the very region above all others where vast beds of coal ought to be found, if vegetation could have produced them. And if they could be found here it would sweep the annular theory from its foundation. The simple truth is that peat and coal are not found where the vegetarian wants to find them. They are found, however, just where the annular theory says they must be found. The vegetarian must leave the very home of vegetation, and in defiance of all law he must find his coal beds amid vegetation stunted and depauperated by cold, and find it, too, where vegetation could never grow,—amid rocks born in fire. Such inconsistencies meet the geologist at every turn. Why is peat found in the ocean? Because the ocean has submerged it. Then why is it found in the thousands of lakes and ponds where no peat vegetation is now growing? Suppose we should find a peat bed 40 feet thick. As it must have been at one time a lake with 40 feet of water, how did the peat begin to grow? Did it begin at the bottom of the lake, and fill the same, or did it begin to grow on the top of the water and gradually drop its carbon particles upon the bottom? In either case it is plain that all rains and floods must have washed mud and other detritus into it far more rapidly than peat could fill it. But there are peat beds from 30 to 40 feet thick. This implies, according to Dana,* 240 to 320 feet of vegetable growth. Such beds were once lakes or ponds, at least from 30 to 40 feet deep. How did such lakes ever become swamps of vegetation without being first "filled up?" Did vegetation fill them with carbon, in order that it might plant itself in a swamp to fill the lake with carbon? This is the pure logic of the peat bog question! Then those deep coal beds of Montchanin,† 100 to 120 feet thick, required vegetable growth of 800 to 960 feet. If that much peat could form and fill a pond 100 feet, a pond 15 or 20 feet stands a fair chance at least.

Now suppose we should find the same regular gradation in the quality of coal in the Southern Hemisphere that we do in the Northern—that is, the heaviest and most massive beds distant from the equator, and the lightest and poorest coals nearer it. It certainly would place the primitive coal theory upon an impregnable rock, even if other evidence failed. Through private correspondence from South America, I have gained enough facts in the case to cause me to place this on record: If geologists will show that such gradation does not exist in the Southern Hemisphere, then the author of the annular theory will take a back seat, where in that event he ought to remain. Here, then, is another test question, with which men under favorable circumstances can either confirm or overthrow my claims.

* "Manual," page 359.

† Phin's "Six Days of Creation," page 64.

There are numberless instances of the formation of bog iron ore, and men suppose that the ore in the bog is the product of a vegetable distillation. But how can vegetation produce iron ores unless it had previously a supply of iron upon which to draw? Are we to understand that the plant *makes* iron? Or that it takes iron which is already supplied and forms a secondary product? We must look back of the plant for all such supplies, just as we must go back of the diatom or the millepore for the matter of silicious and calcareous formations. After these supplies are furnished the builders go to work; and without these primitive supplies there could never have been such formations, either primitive or secondary. On this rock the coal question must stand, and if this primitive supply of carbon had fallen in tropical lands, there would the peat vegetation make its greatest show.

I have elsewhere referred to coal seams among heavy beds of conglomerate—the work of glaciation. Around the Pottsville anthracite region, Pennsylvania, important coal-beds have been opened in the very body of the conglomerate, and must therefore have been formed when the earth lay in ice fetters of a glacial period. The Sharon coal of Pennsylvania, says Lesquereux,* "is placed systematically in the conglomerate," and the very frequent occurrence of conglomerates either directly above or beneath the coal, shows that fuel carbon and continental snows have been frequent associates.

How can the vegetation theory reconcile these inconsistencies? But here is where annular downfalls demand that it should be found, for carbon fuel must have fallen with the frozen vapors, as surely as with other

* "Report of Progress," published 1830, page 630.

annular matter. Coal has been found in fragmentary patches in the silurian beds, and in more extensive beds in the Marcellus shales of the devonian. And the occurrence of bituminous patches of coal in anthracite fields, and the occurrence of heavy anthracite beds, where, according to the old theory, bituminous matter should prevail, necessitates some further consideration in another chapter.

CHAPTER XV.

SOME EMPHATIC AND POSITIVE EVIDENCE OF THE ANNULAR ORIGIN OF COAL IN THE METAMORPHISM OF THE CARBON BEDS;

ALSO,

SOME CONCLUSIVE TESTIMONY FROM THE CRETACEOUS AND TERTIARY COALS.

It is well known that when bituminous or lignitic coal, or even peat, is subjected to a sufficient degree of heat, it is converted into hard coal, or even graphite. Hence, it has become a common belief that all anthracite and other hard forms of carbon found in the earth's crust are metamorphosed beds of soft carbon. Is this deduction a logical one? In immediate contact with volcanic chimneys or where overflowing or intruding lava has heated the adjacent beds, coal has been metamorphosed for a few yards, or even rods; but such heat in various instances, known perhaps to all geologists, has not materially affected such beds except in that immediate neighborhood. How is it, then, that vast coal fields planted in the aqueous crust hundreds of miles from any igneous agencies, except those consequent upon rock pressure, are now in the anthracite, or semi-anthracite, state? If such forms of coal are metamorphic matter, there must be some infallible tests. Let us hunt them up.

In the first place the only purely logical conclusion that can be drawn from igneous intrusion and metamorphism is: That the igneous earth—the great distil-

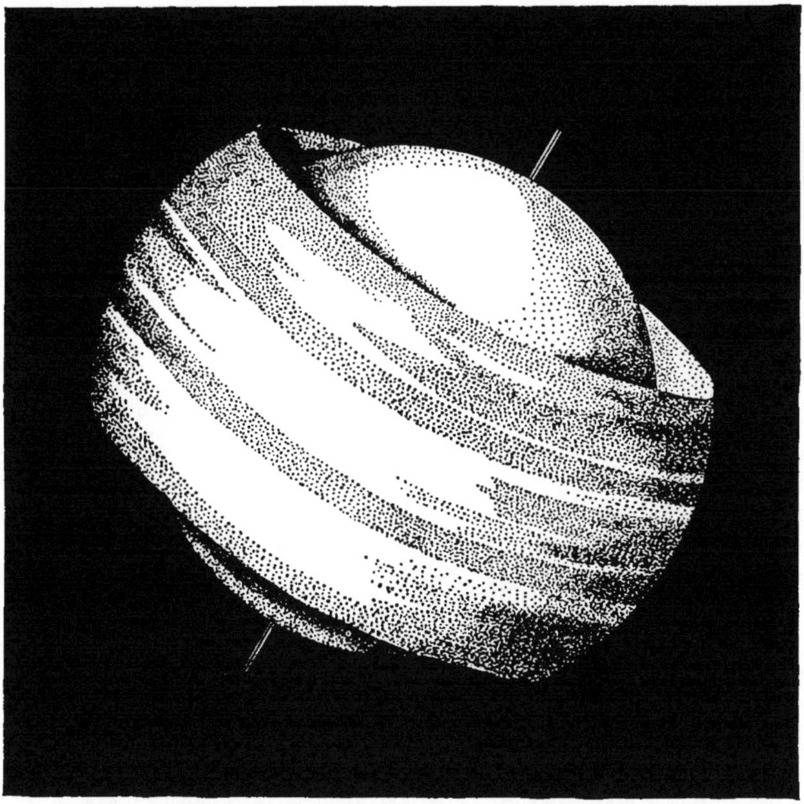

Fig. 12. EARTH IN EDENIC TIMES.

(CANOPUS AND POLAR OPENINGS.)

lery of carbon—must have made its share of anthracite out of the carbon it had control of. If volcanic fires in contact with carbon can produce such beds in a limited way, then the primitive fires of tens of thousands of volcanoes in the incandescent earth, in contact with measureless oceans of carbon, must have made unlimited quantities of the same. Can there be a man of reason, who has ever given the constitution of this orb one attentive and intelligent thought, that does not know that the immensity of carbon now in the earth was in it, or around it, when it rolled through space, a burning orb? Now, if knowing these facts,—these inevitable and self-evident conditions,—the philosopher chooses to utterly ignore them and set up the claim that the last puny fires of a wrinkled and aged world have metamorphosed all this fund of carbon, it shall be no fault of mine. While the peat combustion inevitably points to a previous one, so long as it distils an atom of carbon, the volcano, so long as it changes a carbon bed at the distance of a rod or a foot from it, points to the igneous process, and is proof positive that beds of anthracite and other hard carbons exist in the earth's crust as an inevitable product of that process. If the philosopher cannot see the necessary end to which he is here impelled in spite of education and prejudice I will attempt to persuade him by evidence, if possible, more apparent and conclusive.

Now it is plain that the anthracite coals are either bituminous coals changed by heat to hard coals, or they are themselves an original and normal production; that is, a carbon unchanged, but placed in beds in the form of anthracite.

It is then plainly our next duty first to examine the

evidence of metamorphism, and learn what it declares. When bituminous coal is changed to hard coal it is done by merely driving off the volatile constituents; and as a matter of course, all the ash of a bituminous coal will remain in the anthracite, since it cannot escape as volatile matter. Thus, if in 100 pounds of bituminous coal there were twenty per cent. of ash, or twenty pounds, then by subjecting it to heat sufficient to drive off twenty per cent. of its weight as volatile matter, and thus make a hard coal of it, there would be eighty pounds of coal remaining, including its ash. Now twenty pounds of these eighty are that ash, which instead of being twenty per cent. of coal as before, is now twenty-five per cent. That is, all anthracite coal, changed from bituminous coal, will contain a greater per cent. of ash than the coal from which it is derived. This seems so plain that none surely will attempt to dispute it. It is therefore claimed by geologists that the "average amount of ash in anthracite ought to be one-half greater than in bituminous coal."* Hence it is evident that a fair and candid examination and analysis of coals will settle this question. Anthracite, in order to be a hardened bituminous coal, must contain a greater per cent. of ash.

If, then, during a fair examination we find that it does not contain a greater per cent., then all men will be forced to admit that it never was bituminous coal. After many years of examination, I trust with a spirit of fairness, I might fill many pages with authentic analyses of coals, and hardly an instance at all that I have collected will show a greater per cent. in the average in favor of the anthracite. But I will lay my own tables

* Dana's "Manual," page 363.

aside (partly for want of space), and use only those analyses that all will respect as those of well-known authority. We will first take the analyses given by Dana himself.*

The average per cent. of ash for two anthracites of Pennsylvania is given as 3.46. Comparing this with Pennsylvania semi-anthracites, the former, being more completely changed, as any one can see ought to contain a greater ash. Now the average of eleven samples of these semi-anthracites, taken from the Pennsylvania Geological Survey, and here used by Dana, give an average of 7.16—more than twice as much ash in the softer coal—not very favorable to the old theory, when the harder should contain "more." But lest my readers may think the above average of the anthracite too small to be fair, I will take the average of all Pennsylvania anthracites as here given, and base our calculations upon it. The average of 26 anthracites, including one foreign, is 4.35; but, including only Pennsylvania coals, the average is 5.28. I might use the former, but that the examination may be fair in every way we will use the latter. This, compared with the average of semi-anthracites of the same State, according to the accepted rule, ought to be greater. Eleven analyses of the latter give an average of 7.16, which less 5.28 =1.88, all on the wrong side. Again, comparing it with the six semi-bituminous coals of the same State, used by Dana, we have 10.20 minus 5.28=4.92, still opposed to the rule. Then comparing it with fifteen semi-bituminous coals of Pennsylvania and Maryland, the average of which is 10.32, the anthracite is less by 5.04 per cent. Now if we compare it with the ten an-

* "Manual," page 316.

alyses of Pennsylvania bituminous coals, we have 6.47—5.28=1.19, still contrary to the rule. Comparing with the eleven Virginia bituminous coals, we have 11.06—5.28=5.78. The bituminous has more than twice as much ash as the average of Dana's anthracites. If we go outside of the immediate Appalachian field of coal, and add those of Ohio, Indiana, Illinois and Iowa, we have 398 analyses, with an average of 6.30 per cent. of ash, or 1.02 per cent. more yet than the anthracite. Now as these western coals, as given by Dana, contained a less per cent. of ash than the recent geological surveys give, these tests are very fair. And as any one can see there is not one instance where the rule is vindicated. How it ever happened that this high authority should notice this empirical law, call the attention of his readers to the same, and then immediately complete a large list of analyses, and not see the law completely abrogated, is marvelously strange.

If, now, we leave this authority, and turn to the American Cyclopædia, article "Anthracite," we find eighteen analyses of both American and foreign anthracites, with an average of 4.25, or one per cent. less than Dana's, and a stronger denunciation of the rule. Taking this average I find it to be below the average of the bituminous coals of all parts of the United States, as given by Dr. Peale,* and also below all the western lignites, except those of California. It is not necessary to burden the reader with further statistical facts that may be gathered from both foreign and native coals to prove the utter failure of this alleged law. If one should take the analysis of an anthracite specimen that

* "Amer. Cyclo., Article 'Anthracite.'"

represents a very high percentage of ash he might reasonably conclude that that specimen had at one time been a bituminous coal, and that it was a sample taken from a true metamorphosed region where fire or heat had actually changed it; but taking the whole list of samples thus far analyzed it must be acknowledged by all that the average, so far from being "one-half greater, is at least one-half less." The duty, then, of the geologist is plain: to drop the doctrine of metamorphism in reference to anthracite, and agree that "it is as much a normal creation as the bituminous."* But if he drop this doctrine he must take up the primitive carbon theory. There can be no intermediary doctrine. It requires the agency of excessive heat to make an anthracite, and if not heat brought to bear upon the coal, after it was laid down in the earth's crust, it was brought to bear upon it before it was laid down. And if brought to bear upon it before it was laid, we have no resort but the admission that such coals are the primitive products of the igneous earth. It is only one more instance of the demands of law—one more instance of positive testimony that settles the annular theory upon its immutable foundation. Men who are abundantly better qualified than I am, and with better opportunities for gathering information from the coal literature of the world, can see for themselves that anthracite cannot be a metamorphosed bituminous coal.

But let us examine the anthracite coals under the light of the annular theory. Again, let us suppose a heavy fall of annular carbon in the North Atlantic Ocean; and that the Appalachian Mountains were again under the sea. This carbon carried by the ocean cur-

* "Amer. Cyclo., Article 'Anthracite.'"

rents southward, would fall to the sea bottom in the more quiet waters. The heavy or anthracitic dust in the deep waters would reach bottom where lighter forms could not. Just as a log of wood, not being able to sink very deeply, would float to shallow waters and reach bottom there, while a heavier log, not being able to float into shallow waters, would find bottom in deeper seas. It is plain that if the various coals of the Appalachian field were pulverized into dust, and cast into Baffin's Bay, it would be carried southward along the coast of the American continent, and the light particles would find a resting place nearer the coast. The anthracite dust would settle in deeper basins distant from the shores. That is, the re-arrangement of the carbon would necessarily be very similar to that which is now found in the Pennsylvania coals. The heavy, hard carbon would be planted eastward in the deeper seas, and the soft and light forms would be found farther west. Now I suppose all geologists will further agree with me that before the Appalachian upheaval took place that the eastern base of the system, being farther out in the sea was in deeper waters than the western. How did it then so felicitously happen that the present arrangement of the Appalachian coals and the probable condition of the sea bottom accord with annular arrangement? It is plain that if the anthracite had been placed in the western part of the field, the new theory would here have been a failure. And since we have here three conditions viz., the constitution of the coal itself; its arrangement and assortment in the field; and the condition of the sea bottom—i.e., the sloping from the coast to the deep sea, all pointing harmoniously to the annular origin of these carbon beds, and finally, since we must find primitive anthracite in the aqueous crust, and also thus arranged, what need we of further evidence?

But an objector will ask: Why do the bituminous coals contain a greater percentage of ash? I reply, simply because bituminous dust not being able to settle directly with the anthracite remained longer in suspension, and consequently received a greater amount of marine impurities. This carbon floating shorewards necessarily encountered more detrital matter, and being lighter, settled more slowly, thus allowing more foreign matter to settle with it. Can the vegetation theory in any wise account for this fact? Again, it is plain that if this theory be true, then the farther westward and southward the carbon had to float, the longer was it held in suspension, and consequently the most western and southwestern coals of the Appalachian system must contain the greatest quantity of ash. I need but call the reader's attention to the fact that in the recent Ohio Geological Survey,* of the 200 analyses 45 showed over ten per cent. of ash, 120 give more than six per cent., and 176 exhibited a greater per cent. of ash than the average of American anthracites, as given in the American Cyclopœdia; and the average of the whole list of 200, as given by N. W. Lord, chemist of the survey, is more than seven and a half per cent., or two per cent. greater than the average of ten Pennsylvania bituminous coals, as found in Peale's table of Hayden's report of 1874 (page 177). And one per cent. greater than the ten samples given by Dana (page 316), and two per cent. greater than Peale's average of the anthracites of the Rocky Mountains.

* "Report for 1874," pages 1099 to 1108.

If we compare the Rocky Mountain anthracites with the other coals in the same region * we find there the same evidence, proving that the former were not derived from the latter by metamorphism.

According to this view, then, when the surveys of those regions shall have been completed, the coals in the more southern districts will prove to contain a greater per cent. of ash and the heaviest beds of anthracite will be found planted in the northern part of the great plateau, and principally in British America.†

This necessary rule in the division and assortment of coals is also interestingly illustrated in the northern medial, and more southern anthracites of the Appalachian field. As it is evident that the great inland sea, or bay, in which the carboniferous system of the Alleghenies was laid down, communicated eastwardly and northeastwardly with the ocean, the carbon must have come in from those directions. Then it is also evident that the more eastern and northeastern beds should possess the greatest specific gravity, considering the ash eliminated, while the western and southern beds would be specifically lighter with greater amount of ash to eliminate.

I have but to point my readers to the well-known facts concerning these beds and prove the validity of this position. From hundreds of localities examined there comes but an occasional instance where the facts are not entirely in harmony with this theory. But I

* "Hayden," 1873, page 112.

† Some years after the above conclusion was reached—i. e., that there must be a vast coal field in British America, the following paragraph went the rounds of the press: "A seam of anthracite coal of fine quality has been found on the Canada Pacific Railroad 800 miles west of Winnipeg. The seams are 14 feet thick."

do not ask the reader to depend upon my judgment and observations alone. I will again draw from the American Cyclopædia. In the analysis of Carbondale coal in the extreme northeast coal region, the ash is but 2.70 per cent., the lowest of all analyses given, except one. The next two, one from the Lehigh district, the next important district on the south, afford 2.77 of ash. Still farther southwest in the Pottsville district, forty-two analyses give an average of 4.78 per cent. Still farther south and west in the western district the average is 5.67. Thus taking the various coal fields of Pennsylvania in regular order, from northeast to southwest, and using all the analyses given (except one, exceptionally small), there is a manifest gradation in the amount of ash, which the annular theory imperatively demands, and which the vegetation theory cannot explain.

If this anthracite coal region should extend further south we would certainly expect to find a still further increase of ash; and in reviewing my notes I find there *is* in Southwest Virginia, in the neighborhood of Bush and Price's Mountains, a basin of true anthracite, evidently a prolongation of the Pennsylvania anthracites, where an analysis shows 8.30 per cent. of ash. This may be exceptionally large. But it is not a little remarkable that these analyses should so accord with the requirements of the new theory. Now the density of these carbon-beds is a measure of their specific gravity, and our theory demands that this density should increase inversely, or contrary to the above order, and beginning with the Virginia anthracite and proceeding northeast, we have the following gradation of densities: 1.370, 1.383, 1.510, 1.554, 1.400. Is this all accidental?

I cannot with advantage prolong this argument in the examination of foreign coals, both light and heavy. Enough has here been shown to cause the philosopher to pause and reconsider his conclusions.

Soon after entering the devonian domain we meet with a widespread deposit of black carbonaceous matter, known in different lands by local names, and sometimes divided into two or three divisions, and again combined into one. In many places, as in some parts of Europe, it lies upon a conglomerate and is known as bituminous schists, containing remains of fishes. In various parts of North America it is distinguished by its bituminous or oily character, and claimed by some to be the chief source of the oil flow. But be this as it may, it is a dark or black carbonaceous deposit, and of course it is claimed by geologists that the carbon is of organic origin. Now we have not here the direct and positive means of disproving this claim as we had in the formation of graphite, for here we have organisms, both animal and vegetable, which contain carbon. It is evident that if a deposit is carbonaceous *because* of the presence of vegetable remains, the greater the quantity of vegetation it contains the greater the amount of carbonacous matter it contains. Now nature will vindicate herself. It so happens that these black shales are not nearly so profusely filled with fossils as the rocks either above or below them. So far as my own observation extends the more highly bituminous these rocks are the greater is the paucity of fossils. Dana says:* "The Hamilton black shale is almost destitute of fossils, and very bituminous." Again, in speaking of the Marcellus division of the black shale, he says (p. 271):

* "Manual," page 275.

"The black shales contain but few fossils." But in the face of all this he says (p. 268): "The carbonaceous material of the black Marcellus shale is of organic origin," and also adds that it has not yet been ascertained whether it is due to sea-weeds or land-plants, or partly to fishes or other animals.

Now it so happens that in these shales sometimes fossils are well preserved, and their preservation in abundance in the associated beds proves there was an abundant land vegetation during the time these shales accumulated. There were lycopods, ferns and equiseta, and some of the lower orders of phænogams. There were more than forty species of ferns alone. We find them in abundance in such associations as to show that they are there simply near shore deposits. But in these very places the shales are less bituminous. Now if these plants produced the carbonaceous or bituminous products in the shales, why are not the associated beds, which contain a greater profusion of organic matter, themselves bituminous? These dark deposits are widespread, and it seems impossible to refer them to an organic origin.

Again, when we come higher up in the series we find a great number of dark or black carbonaceous beds, and, as we generally find that these are only expansions and prolongations of coal seams, their origin becomes apparent. Prof. Andrews has said: *"Every stratum of bituminous shale in our productive coal measures implies the existence of a coal marsh on the same proximate horizon, and should always be noted and studied with this fact in mind." He also states that these slates and shales were probably formed out of "carbonaceous

* "Ohio Reports, 1873," Vol. I, page 357.

mud that did not go to form cannel coal, but was floated away by currents and mingled with mineral sediments." I would like the vegetarian to state how the waters carried the carbon away from the marsh, forming on the "same proximate horizon," and first made a cannel deposit, and continuing to float a part of this carbon, and, on the "same proximate horizon," made bituminous shales, without the waters of the seas involving that *whole* horizon. Again, I would like to learn from what proximate horizon the Marcellus shales, and their contemporary carbonaceous beds of other continents, derived their floating carbon.

If the primitive carbon theory be allowed to explain there seems to be no mystery. Carbon that fell in the water and floated directly to its resting place without coming in contact with much detrital matter became a bed of pure coal. The finer particles of carbon-dust meeting with a small amount of foreign or floating particles of clay, would likely form splint coal; a larger amount of clay would form cannel coal, and in the same horizon a part of the carbon meeting with an abundance of other matter would form black slates and shales.

There is another feature in coal that requires a brief notice. I have referred to the well-known fact that fossil plants in coal are generally mineralized charcoal, and difficult of combustion. If the bed were bodily a vegetable production the same difficulty would certainly characterize the mass and we are therefore compelled to admit that the plant is simply a foreign body in a bed of mineral carbon and is itself a mineralized carbon fossil simply because it is in that bed. In short, we are forced to look beyond the plant for the origin of the bed.

Again, mineral charcoal, so frequently found in some coal seams, contains much less percentage of ash than the mineral coal itself. The charcoal frequently analyzes from one to one-and-a-quarter per cent., and sometimes as low as three-quarters of one per cent. That is, the part of a coal seam known to be vegetation is so free from ash as to argue the inconsistency of claiming that the whole bed is a vegetation.

Again, the surface of a coal seam is sometimes covered with undulations akin to ripple-marks. The roof of coal, in some instances in coal mines, plainly indicate that the clays or sand were deposited on a ripple-marked surface, and these undulations are sometimes seen on the face of the coal. These things lead us to conclude that such beds are aqueous sediments.*

To show some of the inequalities and lack of parallelism of coal beds I quote from Prof. Andrews in the Ohio Survey (vol. I., page 352): "In one case within the area of a county where there were five seams of coal in the vertical series the intervals between each two consecutive seams are given. The published figures show that in the subsidence, before the second seam from the bottom was formed, the originally horizontal plane of the bottom seam had sunk to depths

* Though somewhat out of its proper place, I will call the reader's attention to the well-established fact that some of the limestones of the silurian series are distinctly marked with wave-lines or ripple-marks. In the well-known blue limestone beds of the great Cincinnati uplift, are many such undulating layers of wide extent. The mass seems as though it had been thrown into innumerable ridges of from two to six inches high, and when a thin stratum has this peculiarity, the bottom of the furrow thins out, and sometimes disappears. The ridges generally have a uniform direction. These things certainly show that the blue lime strata of the silurian were not deep sea deposits, but actual mechanical sediments, deposited in waters so shallow that waves and currents molded the sea bottom. Hence, the annular origin of these lime beds is apparent.

varying from 34 to 87 feet; before the third seam was formed the second horizontal plane of coal had sunk irregularly to depths varying from 47 to 149 feet; the third plane of coal in turn settled down in some places 31 feet, and in others 69 feet, before the fourth seam was laid down, while the plane of the fourth was found to show an irregular subsidence of from 13 to 40 feet before the fifth and highest marsh appeared with its luxuriant vegetation. . . . If these figures represent facts, they, with all facts, however stubborn, have their rights. *These facts, however, appear to me to have unusual stubbornness.*" (Italics mine.) The Professor goes on to say that such facts are "barely possible," and this must be the conclusion of every philosophic geologist—such irregular subsidence is in fact impossible over such small areas. Again, in alluding to the possible elevation of a coal seam whereby it would form a highland, where rivers might wash out the coal, he says: "So far as my observations go, there is not a shadow of proof of any such upheaval during the progress of the formation of our coal seams, but, on the other hand, all observed facts militate against such a supposition." This illustrious man saw plainly that the swamp theory of coal formation did not admit of such a supposition. But it is plain that such irregular subsidence and re-elevation must produce some highlands. But as highlands did not exist in the coal marshes, as Andrews avers, we have no alternative but to admit that the coal is a sedimentary deposit on the irregular and uneven floor of the sea.

In some places fragments of solid coal have been torn from a seam and carried by currents and deposited but a few feet above the same coal seam from which they were derived, showing that the seam had hardened directly after it had been deposited. Now peat beds, after being deeply buried for ages, remain so soft as to be easily carved with the shovel.

The late surveys in different States have revealed the fact that a coal seam has become so solid as to be planed off as smooth as a board, by eroding agencies, directly after it was laid down, or before heavy beds had accumulated upon it. These things bear witness to the fact that such seams are not vegetable peat formations.

Geologists are all aware that vast beds of carbon were deposited during some of the epochs of the tertiary and cretaceous periods. The cretaceous was the last period of the age of reptilian monsters, and the tertiary the next succeeding period—when the ancestors of our mammalian races came upon the scene. Both periods were characterized by great carbon falls. Extensive coal beds in Asia are probably of the cretaceous period; while the vast carbon beds among the Rocky Mountains, and underlying the vast plains to the east of those mountains, were formed in the tertiary period. I will not crowd these pages with a consideration of these later coal formations, in other continents, but confine our investigation to the so-called lignites of the Cordilleras.

If it could be shown that the great Rocky Mountain plateau, on which the coal beds are planted, did not exist as the sea bottom—over which the waters from the arctic world rolled during the tertiary period—then here the annular theory would meet with a repulse; but it is well known that during that long era of stupendous changes the Rocky Mountain region throughout its entire length and breadth was sleeping in the sea; the ocean's waves then rolled over some of its highest peaks, and the great canyons that interweave that vast region have been made by devouring streams since the coal beds were formed. The tertiary beds reach from Mexico to the Arctic Ocean, proving that currents ran toward the equator along the valley of the Mackenzie, bearing into southern waters whatever fell from the upper world. This, it may be said, is a fact well known to geologists. It is easy, then, to understand how the vast expanse of this western world became the receptacle of tertiary carbon; while the dovetailing of these two facts—that there is tertiary carbon there, and that a great tertiary strait poured its waters southward from the Polar sea—lends strength to the claims I have made. Turning now to the eastern border, and finding no tertiary coals there, we are led to believe that a narrow continent stretched from America to Europe, across the present bed of the Atlantic, thus hindering the southern flow of carbon along the Atlantic sea-board. It is now very fully conceded by geologists that through the tertiary epochs such an isthmus of land reached from Newfoundland to the shores of Europe; if this be true, what a fund of tertiary carbon must lie at the bottom of the North Atlantic! There is every reason for believing that if these later coals had been formed out of vegetation growing in great continental swamps, that the same opportunity was offered by the Eastern sea border for this swamp vegetation. How true, then, the claim formerly made that before a peat-forming vegetation could grow, its foundation bed must first be laid down! Indeed, if there be such a universal tendency for the formation of peat-coal, as geologists must claim, if coal really was formed in that way, the long stretch of coast from Long Island to the Rio Grande presented a great opportunity for the formation of *some* tertiary coal. Why is it not there? The question then presents itself: How did it happen that at the very time our theory necessitated a chance for the flow of carbon from northern waters, an opportunity was given, just where and just when it was required—i.e., over the slope of British America, and when the coal beds were forming? And again, since the absence of these coals on the eastern border of the continent forces the annular theory to demand a barrier across the flood ground from the north, why did this barrier come just where and just when it was needed to support the theory? These, it is true, are minor links of evidences, but they are links none the less.

It must be within the comprehension of every reader that if the vast fund of lignitic coals is a vegetable production, it was present in the tertiary atmosphere as a deadly poison. Now look at the immensity of the coal field at present known to geologists, while every search extends its known limits, and if possible conceive what an atmosphere that was. Turn to the waters of the cretaceous seas and behold them filled with breathing animals, and, if possible, reconcile these facts:—An atmosphere in the highest degree destructive to life; an ocean filled with fishes and reptiles—in many instances there were fishes akin to those of our own time, the recent order of teliosts had ancestral representatives in the cretaceous and tertiary seas; there were sharks like the modern squalodonts, including our salmon and perch; in the same seas was the populous kingdom of reptilian monsters, terrible monarchs of the watery domain, snakes twenty feet in length; the great whale-like zeuglodon, seventy or more feet in length; the shark-like carcharadon, with teeth more than half a foot long, and five inches broad at the base. Turtles lived on the shores and mud-flats, so large that they would outweigh the largest ox of this age. Reptiles of the higher grades, such as the ichthyosauri, combining the several forms of the whale, fish, lizard and crocodile; the plesiosauri, with the body of a porpoise, the flippers of a whale, the neck of a swan and the head of a serpent. These air-breathing animals in vast armies swam the ocean world, with hundreds of other species whose habitat was the waters. But more than all other receptacles of carbonic anhydride—or the very poison the plant required—the ocean, by its wonderful powers of absorption, must have been totally unfit for either fishes or amphibians; and while we know that at the very time the vast deposits of tertiary coals were being made, in addition to the ocean fauna the mammalian types of the present races in mighty hordes possessed the land surface, and which could not have lived in such an atmosphere as the vegetation theory requires—can we possibly reconcile these inconsistencies? A coal-forming age, on a world of abounding life, means the absolute abrogation of law, so long as we admit coal to be the product of the plant, any more than it is to-day.

It is plain that the tertiary earth was in all other respects a perfected world. Fountains leaped from the hillsides; rills and rivers ran to the seas. Birds flew in air; flowers clothed the plain and variegated the forest. Man might have lived then as now. The unwieldy animals, of more than elephantine bulk, argue a heavy atmosphere, which in turn argues upper aerial matter on its way to the earth, which, in fact, is the very thing our theory demands.

Set vast continental marshes to work in the distillation of carbon, how quickly it would exhaust the atmospheric carbon now present as plant food! Then where would the peat-bog look for more to continue the process? From the decay of vegetation of course! But here the millepore fiasco appears again. Vegetation must decay in order that it should produce plant food for vegetation. This we can all see very plainly. But this does not account for the carbon that did not decay —that which is stored away as carbon. Here, then, we are forced to look beyond the plant food in the atmosphere for the deposited carbon. Did it come from the tertiary volcanoes and solfataras as unconsumed carbon? If so it must have used up the free oxygen in the air, and thus have robbed the animal kingdom, which the record denies. Then we are forced to admit that this plant food came during tertiary times, as poisonous carbonic anhydride, directly from the internal fires of the earth. But this means universal death at the very time the world was peopled. It means, too, that the oceans were bodies of acidulated waters, and not all alkaline, as they are to-day; and this means the dissolution of all forms in which lime was a component part, and this the record also denies. What a tribulated path the vegetarian must lead! Besides all these insurmountable difficulties it must have been a very accommodating and felicitous circumstance that the world should produce this plant food at the time so much of its surface was a swamp marsh; that it should fail to feed the swamp vegetation of the tropics, or even that which grew in higher latitudes, provided it was not located where it was chilled and bathed by polar waters.

There are some simple facts which in the philosophic mind must be strong witnesses against the old theory and which the geologist has scarcely yet noticed. Suppose we should take a common limestone and saturate it with coal-oil—such a thing can be readily done under great pressure—we will then have a *bituminous limestone*, just such as exists in great beds low down in the bosom of the earth. If now we subject our saturated stone to great heat we can soon burn out the oil and the limestone will remain solid as before. Not so with the nature-formed bituminous stone. Subject it to the same degree of heat, the oil burns out, but the stone is reduced to impalpable powder. Let us look at this a little more particularly. These limestones thus naturally formed are made up of calcareous particles cemented together by bituminous or asphaltic matter. When there is but a small amount of the bitumen in it the rock is hard and solid, but when it consists of 25 per cent. of the mass, as it frequently does, it is so soft as to be easily carved with a knife. Again, it is sometimes found as pure bitumen in isolated patches or pockets in the body of the limestone. Now as the limestone is an aqueous rock we must conclude that the carbon matter it contains is also.

Not long since a company of Frenchmen successfully impregnated solid limestone with bitumen for asphaltic pavements. But heat would drive out the bitumen and leave the stone hard and solid. Now this seems to conclusively prove that the bitumen was not a cement in this case for, as stated before, when the bituminous stone from the quarry was subjected to heat, the bitumen was also expelled, but the limestone reduced to powder. Does it not prove that in the latter case the bitumen was a cement? It does prove without doubt that the lime particles and the carbon particles were deposited together in the same mass! And as the limestone was matter in the sea, so was the former. Hence, so far as asphaltic carbon is concerned, we see it cannot be a vegetable product since we must look beyond the limestone bed for its origin.

The "pocketed" bitumen is found in stratified seams, and, as both the carbon and the calcareous particles settled together to form the stone, we can readily understand how the assorting power of currents could separate them and form occasional beds of pure bitumen.

Another feature in connection with the anthracites might be mentioned here, although a little out of its proper place. Quoting from the American Cyclopædia will serve to show how utterly baseless the metamorphic theory is: "Prof. H. D. Rogers explains the formation of anthracite by supposing it to be the result of altered bituminous coal, by heat induced subsequent to the formation of the bituminous beds; and he further explains the escape of the volatile portion of the latter as gas through cracks and openings formed by plication. This plication follows closely the general type of the eastern paleozoic rocks which are intensely crushed and folded near the contact of their edges with the igneous or granitic rocks, and much less plicated and contorted in a western direction." But this beautiful picture is badly spoiled by the eminent collators of the Cyclopædia, who, in commenting upon these views, say: "The facts do not sustain the theory! First, the upper beds and strata are more distorted and dislocated than the lower ones, etc. Second, the measures are more plicated and crushed in the western than at the eastern extremity; yet the coal of the latter is a dense hard anthracite, while that of the former is semi-bituminous." It is scarcely necessary for me to add another word on metamorphism, except to say that these facts just stated are precisely what the annular theory claims. If they were otherwise the theory would necessarily fail. A theory that fails in one point is a complete failure. Hence, metamorphism, having failed in the most essential particular, is a complete failure.

Then, briefly summarizing, let us see how the coal problem now must stand in the eyes of the law:

1st. The plant, when subjected to a proper mode of destructive distillation, is made to yield carbon in various allotropic forms. So it is with a limestone, or any other mineral that has carbon as a part of its constitution. The earth was made of such minerals to an enormous extent, and these were subjected to such a destructive distillation during the igneous era; and, therefore, these forms of carbon were placed in the earth's crust, and placed there after the primitive fires died out.

2d. All such primitive distillations existed in the atmosphere of the incandescent earth, which, upon cooling and condensing, formed a part and parcel of the earth's annular system, of meteoric and vaporous matter.

3d. This matter, as it declined and commingled with the true atmosphere of after ages, changed from the ring into the belt-form and over-canopied the earth in its efforts to reach its surface, and consequently fell largely in regions outside of the tropics.

4th. The heaviest form of carbon fell largely in the earliest ages; though all sections of the system must have had some of each form, the outer sections must have possessed the largest part of the light forms of carbon, and the inmost sections the largest part of the heaviest forms.

5th. Thus all ages were more or less characterized by carbon falls, and no age could be exclusively carboniferous.

6th. Carbon, falling into the ocean directly, would separate into lighter and heavier forms and settle accordingly in higher and lower elevations (shallower and deeper parts) of the sea, thus explaining why different forms of coal are found frequently on the same proximate horizon.

7th. The earliest or heaviest forms are free from organic remains, and must, therefore, be a primitive distillation; and the other carbon beds, by their associated strata, by their involved vegetation and other organisms, by accompanying clay-partings, by involved glacial drift, by latitudinal gradation in quantity of ash and specific gravity, by their characteristic absence from the tropics and heavy deposits in higher latitudes, by synchronous formation in all continents, by their evident formation in the very lap and bosom of the glacier—amid ice and flood; by the fact that they are bituminous, oily hydro-carbons, and by a multitude of inconsistencies and impossibilities involved in the vegetation theory, have been shown to be actual sedimentary deposits, and therefore a primitive product.

It has been my lot to pass the greater part of my life among coal veins and coal mines. When a boy I was a coal-digger in my father's coal bank, and my eyes have seen the evidence embodied in this volume, and much more that I cannot use here; and though brought up in the vegetarian school, and a full believer in that doctrine, till forced to denounce it by cumulative and crushing testimony, I must say I have never seen, in the hundreds of coal veins I have carefully examined, one jot of evidence that would lead a philosophic geologist to say it did not evince aqueous deposition. Since then there is not so much as one feature connected with the formation of coal that is not readily explainable by the primitive carbon theory; not one that philosophic law does not resolve into harmony with annular declension, without even the show of conflict; and since vegetarians are forever stumbling upon inexplicable difficulties, boulders, pebbles, pockets, doubling of coal-beds, undulations, slopes, ripple-marks, clay-partings, cannel coal inseparably joined with bituminous coal, anthracites with less amount of ash, marine impurities, carbon planted in archæan beds, air-breathing animals among tertiary coals, carbon dredged from the ocean, dug from the frozen world, and innumerable other objections over which they cannot climb, I am free to say to my brother geologists: Come to this new field! The vegetation theory cannot be true! You all know full well that these stubborn facts are continually multiplying, and before many years roll around you must know full well that normal world-evolution is annular declension. Deny this, and you then must deny primitive igneous action.

It was not until after many years of an effort to explain these difficulties upon the old theory that I consented to connect it with annular matter; but when once placed upon this new foundation every difficulty vanished, and investigators must soon take up the coal question, thus imperfectly treated in these chapters, and bring inexpressible beauty out of confusion.

CHAPTER XVI

OIL, GAS AND OTHER CARBONS.

We now come to the consideration of those compounds which above all others stand as monumental witnesses of the primitive origin of the carbon forms now stored in the earth's crust. It is well known that the bitumen in its rocky matrix, when subjected to sufficient heat, is driven out as an *educt*, not as a product of the rock. Being an educt, we must look beyond the rock for its origin, and where else can we look but back into the igneous earth, which, millions of years before a fish swam the ocean, or a plant rooted in the soil, had entire and complete control of all the carbon in the planet?

When we take the sooty carbon from our chimneys and make it yield oil, asphalt and graphite, it would be difficult for the intelligent chemist to understand how the primitive earth fires could elaborate this planetary carbon into anything else than the very forms now locked in the strata of the world. How can the geologist get around or over this rock, when he knows as well as any one that it was the legitimate business of the primitive heat to make hydro-carbons as we find them to-day? It is not so much a question as to what produced these carbons as to how a molten world constituted as ours is could avoid producing them.

What nature has made in measureless quantities, and is yet, as a puny offspring, of an exhaustless energy, making under favorable and possible conditions, man, with his accumulated and growing knowledge of chemical laws, can make in his laboratory. He can take the living mollusk, or fish, or the human body, or any other organism, animal or vegetable, and subject it to heat in his retort and make all the carbons that were made in the molten earth. But while he can do this, he has no right to conclude that because he can thus manipulate organic matter the carbons and hydro-carbons of the earth were derived from fishes and mollusks. It should not be forgotten that all organic matter is in part an ultimate derivative from and of the molten earth. Reptiles, lobsters and fishes, as well as the plant, are to all intents already hydro-carbons at hand for the chemist to elaborate into anthracite, asphalt and petroleum, and the innumerable other carbon forms. Chemists have made bitumen from fish oil, and paper and shavings of wood answer the same purpose; and he need not stop there, for any organic oil will fill its place. If fishes and plants are to have the credit of making the vast hoards of oil in the earth, why not conclude that the different kinds of fish and different kinds of plants made the different kinds of oil. Of course the olive plant could not make fish oil.

The simple fact that in thousands of laboratories these liquid and solid distillates are being formed to-day by artificial means ought to lead the scientist to conclude that the molten earth, with all the carbon, hydrogen and oxygen right at hand, could not have failed to fill itself with these crude distillates. The grand opportunity thus afforded could not have been shunned, and the chemist to-day redistills what the primitive fire placed within his reach; and this is just what the animal and the plant have been doing all the time in the scheme of world-making. In the artificial manufacture of creosote and carbolic acid man has exhibited but little more skill than Dame Nature. The latter, having had the monopoly in oil making, accidentally, as it were, gave her survivals an opportunity to continue what she had well nigh completed in the igneous age.

The organic oils are not the only source from which hydro-carbons can be obtained. Limestones and all the carbonates, when subjected to dissolving heat and moisture, will yield the asphaltic compounds and oily distillates. If the various forms of bitumen are readily formed from animal and vegetable matter, under a moderate degree of heat, a higher degree of heat will produce the more refractory compounds, as gilsanite, graphite, etc. All the residual compounds, such as remain last in the chemist's retort, are simply these more refractory compounds. Nature left these residuals as the asphalts, graphites, etc. Besides the carbon and hydrogen in the asphalts, many of them contain other elements, as sulphur, nitrogen and mineral ash.

I have in my laboratory a specimen of residual carbon taken from the laurentian beds near Toronto, Canada. It is in a high degree crystalline, bright and glossy, and burns as readily as anthracite, leaving the merest trace of ash. If geologists must insist that this is a product of an ancient vegetation, where not a trace of a plant can be found, when every one must admit that such a product must contain more abundant ash, I can with a thousand times more reason insist that this crystalline fuel is a residual product of the reducing fires that sent oily vapors to the skies millions of years before the earth could support a plant. The thinker is forced to admit that even the plant could never have existed and entered upon its survival work, if these carbons had not been gathered from the earth's inmost depths by reducing heat, and why in the name of reason the plant, as an actual result and product of primitive heat, should be made to do all this antecedent work in addition to its legitimate labors is indeed strange. Why should the plant present the fire-gathered carbon as a *secondary* or inadequate distillation, when, as all must know, the first one was a million times more competent and certain in its work? Why have men so long closed their eyes to the fact that all present energies, all present world-processes, are but *dying* efforts, and can in no sense compare with the Titanic labors of the primeval earth.

There was a day when the world-heat was *beginning*—a day when the reducing fires were starting on their grand career of world-making. In that day the young earth could no more avoid the evolution of the lighter and readily formed hydro-carbons than the artificial furnace can avoid it now. With a heat of less than 200° Cent. fuel gas comes from the chemist's retort to-day, and I assume such tractable products arose from the earth's initial fires, and as sure as fate they went to the skies along with other steaming vapors.

As the earth-fires progressed and the heat became more intense other and heavier products arose, until in the course of eons the world shone as a star, with a glowing ocean of unconsumed fuel. During all that immeasurable lapse of time, let us remember, all these fiery sublimations were being assorted by their affinities and gravital tendencies. The lightest forms would float the highest and arrange themselves in the outermost rings of the system, and as a matter of course they would be the last to descend, and these must be found to-day placed in the uppermost crust of the earth. The heaviest forms of carbon, such as the graphite, would ride the lowest in the fiery envelope and fall early in the geologic past, and must be found in the oldest beds. Is it necessary for me to more than tell the fact that these carbons forms are found just as this orderly scheme demands? The heaviest graphitic masses are met with in the oldest beds only, and the lighter are located away above them.

The heaviest distillations fell back upon a hot but cooling core, and as these all had passed through the fire test everything with them must be incombustible in ordinary heat, and we all know what crucibles of graphite are capable of. Rocks of these oldest formations, as all can see, are not fusible by any ordinary heat simply because they have passed through a higher heat test. Will the old school tell us why and how the readily fused rocks are thus separated from the more refractory? Why have ages separated them? Annular world evolution only can explain it and tell how and why! It will be well to inquire somewhat into the primitive deportment of the oily carbons. During the great oil excitement in the Ohio Valley, when wells were being drilled all along the upper branches of the river, many of these overflowed and ran down into the river channel. When the waters of the river were clear this oil spread over a vast surface, but it was frequently observed that during freshets, when muddy water came down the channel, the oil on the surface of the stream rapidly disappeared. It was found mixed with the fine mud particles—clay, lime, etc.—and sunk to the bottom of the stream. In other words, it was found that petroleum—oily carbon—had at least a mechanical affinity for lime and clay particles. If this affinity exists today then it existed in the igneous period. When clay and lime in measureless quantities went from the world furnace, as sublimated dust or fine mist, into the very region of oily carbons, and if the latter would saturate these particles in the Ohio River, we can see how their combination occurred amid annular conditions.

When the maker of artificial gas puts carbon into his retort, raises the heat, he soon discovers how light hydro-carbon gas readily escapes with a moderate heat, and to enrich this escaping fuel he soon tightens down the valve and injects watery vapor into the retort. This not only enriches his illuminant, but greatly increases the amount of it. Thus we still further familiarize ourselves with the deportment of the hydro-carbons. We learn that if watery vapor improves the quality and increases the quantity of forming hydro-carbons in an artificial furnace, it would do the same thing in the fires of the molten earth.

When the newly-formed hydro-carbons arose to the skies they mingled with the steaming waters formed in the same fires. They not only grew more oily by chemical union, but the quantity was vastly increased, and these oily compounds in turn mingled and mixed mechanically with the fiery mist and dust about them, and when these mechanical mixtures returned to the earth's crust again they became in course of time oil-bearing rock, just such as is now found in the bed of the Ohio river.

Any one can see the necessary deductions from these experiments and practical lessons. We know that there is oil-bearing rock in the crust of the earth, and we know that its quality and its quantity and its position forever exclude the animal and the plant from having any more part in its formation than they now have to-day, and no one that has a particle of regard for testimony will attempt to confer the grand offices of world-making upon them now.

As I write these lines there are ten thousand wells in the United States, each pouring forth from ten to ten hundred barrels of oil per day, and in some places there are from five to ten flowing wells on an acre of ground, and the very thought that organic matter gave this wondrous hoard is too wantonly silly to enter so pure a realm as that of human reason. Then, too, those other amazing oil fields of the old world! All this—and yet the amount now in view is but a trifle compared with that hidden away in lands where the drill has not ventured. Amid the tropic jungles, under seas and oceans, in lands eternally locked in ice and snow, the same fire-formed rocks are filled with this fire-formed fuel.

The intelligent gas maker very well knows that the union of hot carbon with the elements of steam superheated, makes a fuel when locked away from the attacks of atmospheric oxygen, and that the amount of this fuel depends upon the amount of carbon and steam. Then, too, the well-known fact that this fuel was not consumed in its infancy is all the proof we need that the oxygen of the primitive earth found more active affinities in other elements, so that the original hydro-carbons were locked away from its ravages.

The scientist knows how the oxygen must be shut off from his retort in order to form an illuminant, or fuel, and consequently this question faces the old school: How was the oxygen barred from the vegetable carbon? How did the organism decay in the silurian age and escape utter combustion any more than it does to-day?

Did the reader ever see how spontaneous combustion has consumed the piles of "slack" at the openings of coal mines? Did he ever hear of the many disastrous fires occasioned by spontaneous combustion in coal piles and even piles of soot? Oxygen is accountable for these fires, and if it does such things now, after the fuel is formed and locked up, why did it not attend to its offices and complete its work when it had ample opportunity to do so? When a plant or animal dies how often is it sealed away, as in a retort, in order to escape the devourer and become a fuel? This uncertainty diminishes the opportunity to form gas and oil rock a thousand fold; and, as I see it, precious little of it was ever formed as the old school claims. Then, again, in the transition from the oil form to the bitumen or asphaltic compounds and fireproof graphite, how did it escape combustion? I see no possible escape from the conclusion that the fires of the igneous earth formed all the hydro-carbons, sent them to the terrestrial heavens and let them down in world order and covered them up as centuries rolled by. In the world's retort was a vast ocean of superheated steam, as well as an unlimited fund of carbon, and if they did not form a fuel, as the fires of to-day form smoke and soot, then law did not operate. The order of these carbons, as they occupy their places in the super-crust, is an everlasting support of the new theory, and scientists may as well concede the fact now or banish the thought of a molten earth, which they actually do, more or less, when they advocate the claim that petroleum is a species of fish oil. Was the molten earth incompetent to produce the oils out of the material so abundantly on hand at that time? Why, the molten earth could not have made plant food without making fuel! They are inevitable associates now, and they must have been fire-born companions then. Plant food, which made it possible for vegetation to exist, was a collateral product of the same crucible that formed the oceans and the fuel of the world, and it is just as reasonable to claim that the plant formed all the waters of the earth as all the fuel. The animal and the plant are survival products of survival forces in this world scheme of fuel forming, and all they ever did or could do in the way of fuel forming must be a survival product.

Knowing, then, that the first office of the world-fires was the formation of vast oceans of watery vapors, filled with all the mineral sublimations that heat could expel from the globe to the skies, and that chief among these fiery distillates were oily hydro-carbons, we cannot avoid the conclusion that these must be found on and within the earth's crust. Knowing that the oily compounds went into the skies and formed a part of the earth's annular system, we ought to find oil-bearing rock of different ages and at different depths from the surface. Reason ought to lead the thinker to conclude that the oil-bearing rocks and the coal-bearing rocks would be one and the same—all together, if vegetation gave them origin. What, then, must we conclude when we find that such is not the case? The annular student expects to find them in annular order, and when he finds them thus arranged in the earth's crust he willingly leaves the problem in other hands, for the evolution of truth will show who is right.

Oily products sent up by moderate heat from the great world-maker's retort arose much higher in the primeval atmosphere than those which were expelled with raging and excessive heat. The heaviest and most refractory carbon would ride lowest and the lightest would ride highest in the great ocean of fiery sublimations, and in the inevitable formation of world-rings these carbon forms, mixing with their natural associates, would find their level in the annular system and maintain it in all time. Those in the lowest rings would fall first, and as a pure result they would take up their final rest amid the older rocks. Those in the higher rings would fall later and become locked up in the later-formed beds, and oil should be found on all continents.

Now, what are the well-known facts in the case? Oil found on all continents—we might say in almost every land—and new oil fields being continually opened, leads to the conclusion that it is everywhere. Its extent is so great at least as to stultify all claims that it could be of organic origin. Then, again, when we reflect that the hydro-carbons, like all other primitive heat products, fell to the earth from about the polar skies, we would expect to find oil fields under the frozen circles; and if half the reports from those frozen climes be true, this point is also settled. And considering how readily ocean currents can transport such matter toward the equator, we would look to see equatorial beds well filled with hydro-carbons, and such pitch lakes as are found near the equator are witnesses in court. Then, too, we find the greatest oil beds of the world away below the coal beds. The Trenton lime rock, from which such a vast amount of oil now comes in the United States, is located away down amid the silurian beds, among strata that are in no way remarkable for their organic remains.

The formation of oily compounds must have taken place to some extent in the lofty primeval atmosphere, wherever glowing carbon came in contact with superheated steam, which must have existed in all parts of the earth's primitive envelope. In the same region where these changes were going on the appetite of oxygen for calcium, potassium, sodium, iron, etc., so largely robbed the carbon of its natural share of oxygen that the former as a pure result was left as an unconsumed fuel, and as I see it this is just the reason we have so much unburnt carbon in the earth, and I can see no other way to account for the fact that oxygen, ever alert and active, did not consume every atom of carbon in the earth. If the carbon had come in contact with free oxygen, as it does to-day in the atmosphere, there would now be no coal or petroleum to tell the tale, and the old-school geologist might continue to rule the world.

The vigor of free oxygen is shown all the time wherever smoke or unburnt carbon rises from our chimneys. The blackest and densest cloud of smoke from a locomotive or steamboat in a very little while entirely disappears. What becomes of it? Free oxygen has decreased it just as all through the ages gone it devoured the carbon left from decomposing organic matter.

While it is utter folly to look to the mollusks, polyps and fishes of the silurian age as sources of oil, one naturally asks, how could the smoke arising from a molten world, as it lodged amid its mineral vapors and was carried back to the earth and buried as soot a thousand or two thousand feet in the crust, not be made to yield oil and gas? I presume if all the smoke arising to-day from thousands of locomotives could be shut away from the atmosphere and put under as great a pressure as is now put upon the carbon of the Trenton beds, it could be made a source of both oil and gas.

Let us now apply the annular theory in the case of the Trenton rock. It is a good test. The lime which gives a prevailing character to the bed over so great a part of the United States, like the calcareous matter of the cretaceous beds, never came from pre-existing beds. To say that it is an ancient organic deposit is supercilious. The polyp and mollusk might have deposited every atom of it, but they never *made* an atom of it. On the other hand, the lime made the polyp and the mollusk, and we are forced to look to the mineral exhalations of the infant earth for it. There we find it—a vast cloud of calcareous matter floating at its own proper level in the world's great envelope. As it was fire-formed in a molten earth, measureless quantities of smoky, sooty, oily carbon mingled with it. If not, why? A world constituted as this is could not be in a state of igneous fluidity and not send smoky exhalations to the skies; and the fact must be conceded that they floated *somewhere* amid the mineral mist of the evolving planet—better in the Trenton matter than elsewhere, for there we find it to-day. From the very nature of the Trenton matter it became a vehicle for the rising carbon from its volcanic birth. The same fires gave birth to the lime and the hydro-carbons, and as the crystal assumes its form and place, and the plant its habitat, the Trenton matter and its carbon, in order and harmony, dwelt in the great ring-family. Ages rolled away; our ring of Trenton dust, steeped in oily compounds, gradually sank to mother earth, until it reached the outskirts of the atmosphere proper, at the equator. There balancing in mid-heaven, perhaps for centuries, it became an equatorial belt, as all rings in their decline must do. As a matter of course the southern part of that belt was drawn more largely toward the South Pole, while its northern boundary settled more rapidly and largely toward the arctic region. Thus polar lands and polar oceans received the Trenton and its load of carbon first. From these lands the movement was outward upon the vast deep, and the moment they started on their journey other conditions began to operate. Other matter came in contact with the moving matter. The wreck of continents was mingling with the wreck of rings, so that the Trenton deposit was made to vary greatly in different parts, making it difficult of recognition in some parts of the earth. So long as rivers and ocean currents flow, this difference in the same deposit must affect wide areas. Were this not the case I presume the Trenton rock would be a world-wide deposit and oil-bearing in all lands.

When we turn to this deposit in the United States we find it marvelously rich in oil in places and quite barren in others. This fact would seem strange, on the supposition that the lime, as well as the oil, was an animal production, for why should organisms make this vast deposit and yet confine the oil to spots in it? On the supposition that the Trenton matter carried its oily carbon from the skies, it must have given up some of it to demanding currents which was carried elsewhere. To show this assorting power of currents we need only take a survey of the Trenton field:

According to the geologic record, as all will admit,

during the time the vast Trenton bed was forming in the ancient sea, all but the oldest and most elevated parts of North America were submerged, and as a consequence it extends over the greatest part of the basin drained by the Mississippi and its tributaries, and I presume it has its equivalents in all other lands that then were submerged in waters communicating with the polar regions, north and south. The Appalachian system was then unformed, and where they now stand were the abyssal depths of the ocean. In that ocean there was a long elevated and partly submerged region extending from the Canadian highlands to those of eastern continent. Another long and submerged fold extended from Canada southward, and is known to-day as the Cincinnati Arch. This submerged arch divided the deep waters of the Atlantic from those of the present Mississippi Valley.

Let us have this vast body of water in mind communicating with the North Polar Ocean, through such channels as the Baffin's Bay of to-day. Through these channels poured the northern waters as they fell from the skies, and carried whatever loaded them.

In latter times, and yet long before the Appalachian ranges were heaved from the sea, along these great waterways the northworld poured its vast hoards of bituminous and anthracitic carbon into the Atlantic. Antedating all this the Trenton dust, with its associated carbon, had fallen, and as we look over the great field we can trace the currents that carried this carbon and deposited it in favored regions of the sea bottom. Recalling the fact that different degrees of heat sent carbons of different specific gravity to the skies, we recognize how the heavier oily forms floated down into the depths of the sea, and how the lighter, assorted from them in the process of sedimentation, floated to higher regions. In Western Pennsylvania and Eastern Ohio, in the great depths of the ancient sea bed, is the heavy oily carbon, while on the broad top of the Cincinnati Arch the lighter forms were deposited. This is abundantly attested by oil wells of the eastern field, and the wonderful gas wells in the western. The world has long noted the fact, and here it has the explanation, without the assumption that the different kinds of fish made the different kinds of hydro-carbons.

The Cincinnati Arch was a shallow in the Trenton ocean, and I am led to conclude from the very conditions here shown that when the carbon floated down from the north and entered the old Atlantic, it joined a current that carried it westward and southwestward.

It would seem that the great St. Lawrence Valley afforded a channel for such a current. Admitting this, I find many things confirming it. In the midst of such a channel no carbon would be dropped, but along the sides of such a current it would be deposited, and the line of oil or gas wells in New York, from Fredonia westward, and the occasional well on the north side of the channel in Canada, all are testimony supporting the claim of this southwestward oil current. No oil wells, so far as I know, have ever been found in the middle of this old channel, and I presume the cities of Rochester, Buffalo, Erie, Cleveland, Detroit and Chicago, in this old course, will never be able to get oil or gas.

The reader will understand how a wide current moving westward and up the incline of the Cincinnati fold, would drag on the sea bottom and carry all the carbon away from it, just as the prevailing winds carry the leaves up a slope and deposit them in depressions on the top of the hill or upon the opposite side. So the great current moving westward up and over the arch must have left the broad eastern slope mostly barren of oil, and it is well known that it is the rarest thing to strike oil or gas in that region, and the towns of Alliance, Massillon, Akron, Berea, which have spent so much money in the vain attempt to find oil might consult the annular theory to know why. The moving waters carried the sky-fallen carbon away to other fields. But where shall we find it? Certainly we would expect to find it on the broad top of the arch in the depressions and valleys that run over and along, but most abundantly on the western slope of the arch. Of course all the world knows what a remarkable oil and gas field both these regions are. The great "Karg well," at Findlay, Ohio, was drilled into a depression on the very summit of the arch, and a large number of wells in that vicinity, and also at Fostoria and other places, have made that region one of the most renowned oil and gas fields known. Farther south is the famous oil field of Lima. All these are on the broad fold of the Cincinnati Arch and on its western slope.

These well-known oil fields make the annular student bold to further hypothecate. From the great St. Lawrence current a branch current seems to have run south into Pennsylvania and the Ohio Valley. In making this course it would have to pass over a long ridge in the old ocean, running from southwestern New York, from Chautauqua Lake southwestward, and when we come to view that magnificent if not peerless oil field of the Pittsburg region, on the southern slope of that ridge, we are still more inclined to think this hypothesis a correct one. And when we further find two lines of oil wells along the Ohio Valley—one on either side of it—we add another link to the chain of testimony. And yet another link is added by the fact that the oil wells of this valley are mostly located on the southern slope of the Ohio hills. I have found this to be the case so often that in consultation with oil seekers I have always advised boring in a region sloping southward or southwestward and have been gratified to realize success, while I have seldom known success elsewhere in the Ohio Valley region.

The question may now be asked, Why do oil fields *run in lines?* Oil men and geologists know very well that this is a peculiar fact. They first get the direction or trend of an oil field and then drill succeeding wells along that course, and even locate "side lines" or branches and follow these in course. Those were peculiar fishes that died and decayed by line and plummet.

The great St. Lawrence current ran across the northwest corner of Ohio, along southern Michigan, and it is not a little strange that in passing northward from the Findlay gas region we come into a region utterly barren of oil and gas. Plainly we cross the *middle* of the old current's track, in which no oil wells will likely ever be found. Farther north, toward central Michigan, the north line of the current *may* be found. Time will decide the case.

When we comprehend the fact that thousands of oil and gas wells are yielding millions of barrels of oil and countless millions cubic feet of gas, year after year, until generations pass and leave them to their children, we are forced to marvel that vegetarians still enjoy the majestic scene.

Why is the Trenton bed so barren of oil all along the eastern slope of the Cincinnati fold? Why a field so amazingly rich on its broad summit and western slope? What kind of fish and crustaceans could those ancient oil- and gas-makers have been any way?

And now a little history, however it may show the egotism of the writer. It is not difficult to recall the oil excitement in the seventies and eighties—how thousands of men with millions of money ran wild in an effort to discover new oil fields. The author of the annular theory and a few of its advocates urged moneyed men to drill into the Trenton rock for the hoard it held. The vast fund of carbon below it in the older beds stoutly affirmed that a lighter oily carbon must be located above the heavy graphitic carbons. But geologists everywhere, I believe, discouraged the attempt and denounced the claim. Failure to find oil on the eastern slope of the Cincinnati anticlinal confirmed them, and it was not until about the year 1885 that a few prospectors ventured to bore on the top of the arch in northwestern Ohio. I remember how our opposers called us "crazy," "cranks," etc., and how it was announced from official position that "No oil can be found in the Trenton and no gas beyond the Maumee River." But the drill was put to work almost on the very summit of the "arch" and kept to work under discouraging conditions until gas gushed from the well with terrific force. The history of that "mighty gusher" is well known, and I have no room for it here. One year from that time the town of Findlay began to put on city habiliments, and resolved to have an anniversary in commemoration of her "Application of Natural Gas to the Mechanical Arts." A gas carnival and banquet was held in the young city, and it was estimated that fifty thousand people assembled to hear the orator of the day, the great "Karg well," whose mouth was opened in defence of the annular theory, and the very earth trembled as it spoke. The Ohio State geologist was invited to be there and tell why gas "could not be found in the Trenton nor beyond the Maumee." The author of the annular theory was invited to attend and tell why that bed was filled in places with oily hydro-carbons and gas. The former was not there, but the latter was present and spoke according to program. It was the proudest moment of his life when he finished his lecture, and the "Karg" closed the midnight banquet with its terrific roar of approval.

In that lecture the speaker made some predictions, which have been fulfilled, and here they are:

"I tell you, ladies and gentlemen, the great Karg gusher is not the last one to be found in this field. West of Findlay, to the very bounds of the State, and on into Indiana and perhaps Illinois, the western slope of the Cincinnati fold extends, and it is laden—it must be laden—with oil and gas.

"On the northern borders of the State, and in southern Michigan, was about the center of the great St. Lawrence current. You need not drill for oil in that region. All the hydro-carbon was carried away by the rapid waters.

"The branch current of the Ohio Valley, upon meeting the deep sea in that region, west of the arch, would have its motion checked, and perhaps an eddy was formed where now is southern Indiana, and I predict that a great oil field will be developed there."

As these predictions were made long before the Indiana oil fields were heard of—before it was discovered that there was a barren line running from Lake Erie on the southern boundary of Michigan—they become interesting links of annular testimony.

This little episode might close this chapter, but I have before me the vast oil regions of the Eastern Continent, from whose beds the oily floods have been pouring and burning as sacred fires since the night-time of history. Who can estimate the vast sea that has been escaping from away back in the ages? China, Japan, Persia, Russia, North and South America each pours a river of oil into the world's trade and I presume will do so for centuries to come, and when the known oil regions fail we may boast that other and vaster fields remain untouched—and must I believe that the fish and mollusk made all this? Are they making oil to-day? If so, where? If not, why *then* and not *now?*

CHAPTER XVII.

CONCLUSIVE EVIDENCE OF ANNULAR DOWNFALLS IN THE TERTIARY OCEAN OF THE NORTHERN HEMISPHERE.

Over an extensive portion of the Rocky Mountain region the tertiary beds, as might be expected, are fresh water deposits. During the cretaceous age, as is well known, this vast area was covered by the sea, and these waters had communication on the north with the Arctic Ocean, probably by way of the present depression in British America, along the valley of the Mackenzie River; while on the south it communicated with the Gulf of Mexico, or other southern waters, by way of the lower Mississippi Valley. Thus a wide channel or strait passing from the Arctic Ocean fed the waters of the great cretaceous sea of the Rocky Mountains, and through this sea had direct communication with salt waters of the south.

Now, if my claim be a valid one the beds in the Rocky Mountain tertiary will present the following features: The cretaceous period having been brought to a close by a down-rush of waters and snows in the Northern Hemisphere, a stream of water pouring southward through the above named channel must to a great extent have been a fresh-water current; and those deposits in the extreme northern beds in the area under consideration must be in a great measure fresh-water accumulations. Those in the middle part of this region must be fresh beds to a less extent, perhaps sometimes marine and sometimes entirely fresh, owing to changes in currents, etc., and here fresh-water and marine species will be commingled. While in the southern part the beds must be almost exclusively marine. This conclusion I came to long before I examined the records. It is the conclusion which any one familiar with the manner in which fresh-water and salt-water currents of the seas dispose of living organisms will come to. And further, we will reasonably expect that on the Pacific and Atlantic coasts, where tertiary beds are found in the same latitude, and where the open seas have access to the shores, marine fossils will prevail. It must be seen, then, that these conclusions are based upon the fact that a stupendous addition to the oceans in the closing cretaceous came via the channels from the polar sea. Now if such things are not to be found our theory must receive a stunning blow. Let the reader reflect for a moment on the philosophic distribution of oceanic life under such promoting causes; and we will then endeavor to learn something of the character of the tertiary beds of the Rocky Mountains.

Dana says: * "The general distribution of the marine beds is similar to that of the cretaceous . . .the inner limit being about 100 miles from the Gulf in Alabama, 150 to 200 in Texas, and along the Mississippi River the Gulf border extends northward to southern Illinois." Again, "the fresh-water or lake deposits are, as stated, of all periods from the middle eocene to the pliocene, the eocene occurring about Fort Bridger, the miocene in the upper Missouri region." "It occurs, also, in the Big Horn region, in Chetish Mountains, about Fort Union." "It extends far north into British America, and south to Fort Clark and beyond to Texas." And now mark (italics mine), "In the *lower*

* "Manual," 490 to 493.

part, on Judith River, there are *brackish water deposits*, containing shells of *oysters mingled with fresh-water shells*."

Again, "In the Rocky Mountain region the lignitic group of the Green River basin, near Fort Bridger, etc., consists of sandy beds, some of them true marine, more of them having a commingling of fresh-water shells with marine, which indicates very shallow brackish waters, and a still larger part strictly fresh-water in origin."

Thus there were conditions by which brackish water-beds were formed in the southern part of the tertiary sea, on the Rocky Mountains and almost exclusively fresh-water strata on the upper Missouri. While in the extreme south, as in Texas, the beds are wholly marine, and in the extreme north wholly fresh-water. For it is well known that there is no marine tertiary in the latter.* But this condition was strikingly different from that which immediately preceded it. Says Dana:† "In the closing part of the cretaceous, in the Rocky Mountains, there was a change permanently from a condition of general submergence under salt-water to one of oscillation," etc.

Thus we see along a wide strait running from far north, southward, came in a fund of waters unsuited to marine life. It was a supply of fresh-water that was of sufficient volume to drive marine forms southward! If this fresh-water came from rivers, where did they rise, and whither did they flow? Did some river flow from the north, or from the east, and empty into a sea forty times as large as Lake Erie, making its northern

---

* See Dana's "Manual," page 488.

† Ibid., page 478.

part fresh, its central brackish and its southern marine? It is not likely. So vast an area of fresh-water formations cannot be explained by lakes or lacustrine deposits. It is evidently the work of a vast bay fed on one side by a fresh-water ocean, on the other communicating with the salt ocean.

Here is a distribution of fossils that gives valuable support to the claim I have made. A new fund of fresh-water came in at the beginning of the eocene, when its waters were filled with marine forms; gradually but surely these gave way, so that before the formation was half completed fresh waters had so gained upon the marine that fresh-water formations are reckoned from the middle of the eocene. Marine forms are pushed southward into marine waters, while about midway between the two extremes the fresh-water and marine are so commingled as to be sometimes fresh and sometimes salt, favoring neither true marine nor fresh-water organisms. Now I suppose there is not a geologist living who, upon examining these things, will not claim that the fresh-water came from the north; and I am sure he will not claim that it was river-water.

Now what means this peculiar arrangement of strata? Could they be more emphatic in their testimony to the truth of a great fresh-water polar sea if they had been intentionally arranged to lend it support? Suppose the tertiary of British America from the United States border, along the line of the Mackenzie to the Arctic Ocean had been marine. It would have been a crushing evidence against the annular theory. But we are not yet through with this investigation. Dana, our same high authority, says: "The tertiary of the Pacific coast is of marine origin," * and also that the marine tertiary covers a large part of the Atlantic border. Doubtless Davis Strait at the same time poured a volume of fresh-water from the polar world directly into the Atlantic close to the North American coast, just as the channel of the Mackenzie. For we find the same commingling of marine and fresh-water fauna on the New England coast, while in the northern part the shells are exclusively fresh-water species. We are not at liberty to call these river and estuary deposits, for all the estuary and river deposits farther south, on the sea border, are chiefly marine.

Along the coast, from Delaware Bay to Florida, and around the Gulf of Mexico, where hundreds of rivers, including the Mississippi, empty wherever the tertiary beds are laid down they are not considered fresh-water beds. Hence the utter fallacy of the claim that the vast expanse of north tertiary beds are fresh-water lake deposits, or of fluviatile formation. How could it be possible that, in the absence of important rivers, such wide reaches of fresh-water tertiary could have the origin claimed by geologists, when all the evidence is that the mightiest rivers of the world pouring into the ocean have failed universally to make such? Now, if we will just conceive that the vast polar ocean of the tertiary period was a body of fresh-water all mystery ends.

Thus, on the California coast on one side and the New Jersey shore on the other, we find marine beds deposited synchronically with the fresh water deposits of the tertiary sea of the interior, showing that in these parts of the earth, where the open sea or oceans washed the shores, salt-water prevailed. But here were lands

---

* "Manual," page 492.

washed by river streams! Virginia and Maryland, skirted by marine tertiary, as is well known, had rivers pouring into the sea, for the Appalachian arches had been previously formed. Why did not these rivers cause fresh-water deposits, or at least brackish beds? The simple fact is, that the very tertiary formations which we well know must have been somewhat under the influence of river water, do not show such an influence, while in the interior tertiary sea, where, so far as we can tell, no rivers emptied, the whole deposit, from the National Park to the Arctic Sea, is a fresh-water formation. Hence the reasonable conclusion that my claim is a just one. *T*hat so-called lacustrine, or fresh-water deposits, in the Northern Hemisphere are not necessarily river-formed beds, nor deposits in inland seas, as is generally claimed.

But the picture is not quite complete. Something is needed to finish the triplicity of phenomena. We have had the "plunge bath," and the extermination of species, which Dana says * is "Remarkable for its universality and thoroughness." New waters were poured into the oceans; therefore, as greater mechanical pressure necessarily resulted, and consequent increased heat and expansion in the deep-seated rocks beneath the ocean's bed, we must look for crust folding and upheaval. And as this downfall of water closing the cretaceous period was a stupendous one, and the extermination "universal and thorough," then the crumpling must have been correspondingly stupendous. Now, what do we find? Early in the tertiary we find mountain making on every continent—a grand world-wide disturbance of strata, equaled only in its universality by

* "Manual," page 488.

one of the most "complete exterminations of species of which there is record." * In a former chapter I have shown why this order is so invariably maintained.

Now, here it will be seen that all these phenomena combine to demonstrate the truth of the annular theory. A downfall of water necessitates accompanying snows and a change of climate; hence the world-wide extinction of life-forms well-known to all geologists, and since I know of no competent cause of universal strata-folding, but increased mechanical pressure universally upon the ocean's bed, I am simply forced to the conclusion that the cretaceous period was closed by a stupendous downfall of tellurio-cosmic matter upon the earth.

Now, it may be somewhat interesting to look a little into these changes, as recorded by the races entombed in the debris of continents. At Jackson, Miss., the eocene beds contain numerous marine shells, and here have also been found the giant remains of the zeuglodon, a whale-like inhabitant of the cretaceous seas. In the Green River basin are found the remains of fossil fish belonging to the cretaceous waters and buried in the early tertiary beds; also mammalians of the tapir family, and remains of the dinoceros and uintatherium.

While on the Atlantic border, from Martha's Vineyard to southern Virginia, cretaceous animals in great numbers are found in marl-pits of the lower tertiary. These things seem to confirm the claim that the cretaceous world was swept by a mighty cataclysmic wave, and that its animals were buried in the detrital mass swept from the land into the seas and which formed the lower eocene beds.

* Dana's "Manual," page 487.

So nearly are the lower eocene beds related to the cretaceous that eminent geologists are unable to agree as to whether they are tertiary or earlier. Dana says: "It is doubtful whether they are one or the other." Both Profs. Cope and Marsh discovered in eocene beds remains of saurians, related to the dinosauri and megalosauri, which are known to have been cretaceous forms. Thus so far we see that the "American Record" shows that new waters brought in a new environment, involving a general destruction of cretaceous forms, and buried them in the debris of the cretaceous world, carried in a great revulsion to the seas, which became the lower tertiary beds—actual transition beds; hence the difficulty in assigning it its true place in the series.

Again, when we turn to the foreign tertiary we find the same general conditions prevailing; especially after the eocene, or first tertiary, there is a general prevalence of fresh-water beds over a large part of Europe. Let us briefly examine these things and see how they bear upon a fresh-water ocean.

In a recent lecture by Boyd Dawkins, the English geologist, while speaking of the former conditions of the earth, in more modern geological times, he said: "In the eocene and miocene periods Europe was united with Iceland and Greenland, and also with the United States of America, by a barrier of land extending past the Faroe Isles, which was covered by a dense forest, composed to a large extent of the same trees as in Europe and in America, and which allowed of a comparatively free migration of animals to and fro between England and the United States."

I suppose the conclusion was drawn from the similarity of the fauna on the two continents—the reptiles and the fishes that formerly inhabited the streams and lakes of both continents. Now, while I am free to admit the existence of land communication between Europe and America by way of that great submarine plateau, I am not able to draw the same conclusion from the evidence, since I cannot see why land communication is necessary for fishes and alligators to pass from one continent to another, unless a fresh-water channel or river ran from one extremity of the plateau to the other.

But if we now admit, as I have before urged the necessary fact, that the incessant fall of exterior vapors in the northern ocean—the measureless fund of snows deposited and melted in its waters—produced a great fresh-water ocean, that involved the north polar world; and that, with the exception of a few channels that connected with southern waters, it washed a continuous barrier that encircled the earth, the explanation becomes easy. It is claimed that during the miocene period the climate was tropical, even where now the winters are severe. Suppose, then, that during the miocene, or earlier in the tertiary period, the Mackenzie flowed as it now does, and was inhabited by fishes and other animals of a tropical climate, and that other rivers ran from the great plateau northward into the same sea, just as the rivers of northern Europe and Siberia do to-day (and it would be impossible that such streams should not exist), it can be readily seen, according to this hypothesis, how the same fauna that characterized the land of the Mackenzie would characterize the entire land belt of North America, Europe and Asia.

These continents were all washed by the same north sea. They all poured mighty rivers into the same. And the waters of such a sea were fresh under the same laws that to-day make Lake Superior a fresh-water sea. There would be very natural facility for an intermingling of species, and I presume such means are much more reasonable than to suppose a great isthmus of land, for it must be remembered that there were likely straits connecting this north ocean with the waters of the Atlantic, since in a part of the tertiary times fresh-water and brackish-water fauna, driven from the north, inhabited New England seas, and many mammalian land animals are of different species. The mammoth of Europe and Asia was different from the American; but the hypothesis of a fresh-water ocean does not rest alone on this kind of evidence. The widespread fresh waters of the tertiary period do not apparently admit of any other hypothesis. One-third of North America, a great part of north Europe, and very nearly all of Siberia, and much of China and other parts of Asia, were apparently synchronously submerged beneath fresh waters. And it certainly would not be too strong language for me to say they were submerged by the north polar ocean.

Geologists have long claimed that the great fresh-water beds of tertiary Europe were made by great rivers, running south or southeastwardly from a northern or northwestern continent. What, then, deposited the tertiary beds of Siberia and North America? If this claim be true, what are now the continents were then the oceans. But the evidence is accumulative that an elevated arch of land, once formed in the evolution of continents, always remained an arch; and we have no evidence that such ever became a trough of the ocean. Besides, the geological record of the tertiary itself does not corroborate this view. The filled up estuaries show that the same rivers ran into the northern ocean that now empty into it, only at a higher level, and the fossils of alligators and other inhabitants of rivers show that the land was only partially submerged. If the rivers ran in the opposite direction analogy would show that these great fresh-water beds could not have been such as they are—so exclusively fresh-water deposits. It seems to me, as we look over the vast field, it is impossible not to be convinced that a great fresh-water ocean rolled its billows over the so-called estuary beds of northern Europe, Asia and parts of North America. It seems like reversing natural tendencies to conclude otherwise, and I am sure when the sober calculation of man is brought to bear upon this great question it must gravitate into the line I have here indicated. With this thought before us a hundred mysteries are explained. The fresh-water beds of Norway and northern Russia, those of England and Scotland, can then be explained; for being identical in many respects with those of France, that they have all been supplied by the same northwestern continent, would seem unreasonable if not impossible.

But now, with all the evidence of a downfall of vapors, as shown in North American tertiary beds—in the grand slaughter of living forms and the folding of strata, with all the evidence of a nearly isolated ocean as against the evidence of a northwestern continent in the formation of beds on three vast continents, let the reader say which is more in accord with law in the evolution of the earth. If we leave the continents standing, with their outlines somewhat contracted by necessary submergence, with the same drainage system they now possess, and still further admit the necessary fresh-water ocean to account for the rock-recorded history, I believe the interpretation becomes plain.

But what does this great ocean of fresh water prove? Does it not point with almost positive conclusiveness to an augmentation of snows from the great super-aerial fund? Have we not almost positive testimony in abundance that the cretaceous age just closed was ended by excessive and universal refrigeration? That the transported blocks of stone found in the upper cretaceous and lower tertiary point to a northern origin? These things being apparently true, we are again forced to admit that in addition to the reasonableness of such a conclusion the evidence is overwhelmingly in favor of an annular fall of waters in the north polar world as the sole cause of the transition from the cretaceous to the tertiary.

Let the reader now draw somewhat on his imagination. It is well-known that all the existing continents were largely submerged under cretaceous waters. The Rocky Mountains, the Andes, the Alps and the Himalayas, were either unborn or in their infant stages. It was a universal ocean of calcareo-saline water. Then it is evident that some mighty barrier was reared by some resistless force that rolled the cretaceous waves southward and made an isolated fresh-water ocean in the north. But here is our hypothetical barrier, the great Atlantic plateau, reaching from the coast of Newfoundland to Ireland, and known by actual soundings and other evidence to be a table-land submerged. It was raised from the deep, according to the record, at this very time and stood for uncounted millenniums as dry land. That is, we have an actual barrier at the very time the fresh-water ocean was formed. We have the three-fold phenomena of new waters from on high, wide extermination and plication of strata. As I have before said, there is enough water on the earth's surface to-day to make one thousand terrific cataclysms, each of which would cover the whole earth fifteen feet deep with water. Suppose now a fall of snow in the northern regions sufficient to spread that amount of water over the earth packed into glacier ice. The actual mechanical pressure, incalculable and inconceivable, arising from such an additional amount of exotic matter in the polar world, forcing the lower rock-beds into a condition of plasticity, must have had every pound thereof conserved in crust upthrowal.

A vast mass of rock moving in obedience to a measureless directing power (just as a glacier on the earth's surface moves under Titanic pressure), is simply forced under the plateau, and this, with the additional force of rock expansion under augmented heat, in the inmost depths of the earth, which no resistance could curb, gradually, but with a step as steady as time, raised the plateau until the towering ice continent, at one end of the telluric balance-beam, was equipoised by a new and growing continent at the other.

Now, we know that this new land was raised by a force directed at right angles to its axial line running east and west. The southern force was evidently a resisting or passive force; the northern an active energy. We also know that a force working thus lifted a belt of land reaching from western Europe to eastern Asia at nearly the same time. What, then, must have been the volume of that lifting force? No wonder a new continent was made. Suppose an ice cap five thousand feet thick should suddenly cover the Arctic world. What would the pressure of such an ice-continent result in? Is it not physically certain that it would press that part of the earth inward or downward upon itself, even though the planet were solid to the center? Sixty thousand feet of steel blocks piled one upon another would give sufficient pressure to render the lowest blocks plastic, and raise the temperature thereof nearly to the point of fusion. Suppose a mer de glace were placed in the bosom of the earth and covered by thousands of feet of rock. It is evident that if a greater vertical pressure than lateral were exerted the ice would move laterally until the two forces became equal; and it is easy to conceive a vertical force sufficiently great to press the entire mass laterally into another bed. It would be just so with a granite bed or stratum of steel.

A downfall of annular matter must add additional pressure to rocks, perhaps already yielding to a directing force; add more heat and consequent expansion, which no terrestrial resistance can withstand, and a continent, it would seem, rises because additional matter, by a resistless force, is intercalated between its surface and its foundation beds.

Let us remember that as the tertiary was a time of great mountain making and consequent changes in sea-level, those portions of the continent which necessarily lay near the regions of disturbance were subject to oscillation by depression and elevation of barriers, so that inlets from southern waters, reaching into or near the fresh waters, sometimes commingled their fauna, so that, as in the Paris basin, salt water sometimes occupied the ground, and again fresh water returned, and that these conditions may even have extended at times far into the British Islands. One fact, however, is very plain, that after the eocene the northern portion of the continents, both eastern and western, were seldom washed to any extent by marine waters. The evidence seems to point to the fact that mountain making in Europe was a very irregular process. The Pyrenees, likely, were elevated during the eocene as well as the Julian Alps, and some other highlands. But the Apennines delayed until the close of the eocene. The Western Alps, where stands Mount Blanc, arose at the close of the miocene. Some of the British mountain chains were likely elevated nearly the same time; while there is also evidence that mountain making in Central and Southern Europe was active throughout most of the pliocene, showing that the tertiary was eminently a period of disturbance.

Further, let us remember that if a tree grown at the headwaters of the Mackenzie were to float into the polar sea, as some doubtless do to-day, it would be no evidence when found buried in the polar sea beds that a climate of the Upper Mackenzie prevailed beneath the Arctic circle. And when geologists of to-day find the California pine in the miocene of Greenland, or the cypress of Arkansas in the miocene of Alaska, or Spitzbergen, they have no right to claim a warm or subtropical climate for those regions on this evidence alone. For when we assume an elevated plateau across the Atlantic we must also assume river systems, drawing the same into northern waters, and the transfer of southern plants to northern beds; so that much allowance must be made in the claim that a subtropical climate has existed in those lands when based on such evidence.

Now, as we look back over the tertiary world, and are advised from its well-known record, I cannot see how a geologist can come to any other conclusion than that which here is evidently forced upon us. We cannot shut our eyes to the overwhelming evidence that shows one vast expanse of fresh waters. 'Tis not in the Paris basin alone, not the whole of northwestern Europe alone, nor the stretch of thousands of miles on the northern coast of a single continent, that presents this testimony. It can scarcely be possible that a fresh-water lake of one-fourth the expanse of the tertiary fresh-water beds of Europe could obtain. But when we can trace the shore lines of this limitless fresh-water sea around the whole hemisphere, we are driven, it seems to me, beyond the possibility of a doubt, to the conclusion that during the greater part of the tertiary period the great Arctic Ocean was a wide expanse of fresh waters.

But what does such a conclusion lead to? It leads directly to the positive and permanent establishment of the annular theory. A fresh-water polar ocean, coming immediately after the cretaceous period, means a vast down-flow of annular waters or snows—the very thing demanded by the cretaceous glaciers, the very thing demanded by the elevated barrier, the very thing demanded by the sweeping and universal extermination of species. If no other evidence could be found to support this theory a polar fresh-water ocean rolling as it were over the beds of cretaceous matter would seem to settle the question beyond a doubt, since no other terrestrial source or cause can be found.

Thus it seems that every step we have taken in this long and to some, perhaps, tedious investigation has added a link of testimony in favor of the grand conception of an annular system. We read the thought on the gilded firmament—the clock-work of the heavens; we read it in the solid rock-ribbed earth traced in imperishable lines, from the close of the archæan time till the last great fall of waters—a thousand links joined and interjoined—a multitude of witnesses speak from every field. Our knowledge of the earth is yet exceedingly limited. Geology is yet in its infancy. Man is just waking up and laying hold of the great volume. And the earth—conceived in nebulous heat, born in the throes of the mightiest revulsions, rocked in the billows of a molten sea, and swaddled by its inveterate flames—grew old and now treads its majestic round, clad in the wreck of rings, its bosom filled with the dust of races. From that dust man has arisen, and looks back upon the bed whence he came with bewildered eyes, and forth upon the possibilities in visions bright with Hope.

APPENDIX.

NOTE I.

THE LAST ADVANCE OF GLACIERS.

It must seem plain to the reasoning mind that if the last downfall of exterior vapors fell at the time of Noah, and was, as is here claimed, the deluge, then there must have been a modern advance of polar glaciers, agreeably to a physical law referred to before.

As previously stated, when exterior vapors entered the atmosphere, they gravitated naturally more largely toward the polar regions, and, falling there as snows, would accumulate there as glaciers, and the extent of those glaciers would correspond to the amount of snows. Now, it is evident, if there ever was an Eden climate on this earth, its destruction was brought about by a change of climate. It is also evident, if the deluge was a collapse of the last remnants of upper waters, that the latter must have begun to fall in polar regions many centuries previous, since we see that throughout all geologic times such changes are spread over vast periods.

It appears that the Eden world suffered a change during the Adamite age (Gen. 3: 17 to 22; also Gen. 4: 12), and it also appears that that change was effected by a change in climate. For the race which dwelt naked in Eden became clothed in the skins of animals; and whatever interpretation the opinionated may draw, I draw my conclusions from law. That, if the infant human race ever dwelt naked on earth, the climate was then warm, and if it afterwards dwelt on earth clothed in the skins of animals, it had then become colder. And if it grew colder, it is more than probable it was caused by a fall of snows; and if Eden was formed by means pointed out in these pages, then it must be almost certain that the Edenic climate was changed by a fall of snows from the earth's annular system.*

---

* See "Eden's Flaming Sword" in the author's second volume on the annular theory.

Hence we have reasonable grounds for concluding that for more than a thousand years of the Edenic period the vapors which finally involved the earth in a terrific and wide-desolating flood, continued to fall as snows at the poles. And, if the physical conditions of the antediluvians and their environment depended upon the conditions of upper vapors, of which there is no reasonable doubt, then the question is almost reduced to a demonstration that polar glaciers began to advance in Edenic times; and thus we have connected into one grand and varied scene, the whole age of antediluvian man. The same cause that deprived him of his Eden home, and brought upon him all of life's physical ills in a modified form, closed the scene at the time the "fountains" of the aerial "deep" were "broken up."

Now, if this be true, there must be some physical evidence of a change in climate. Let us briefly turn our attention to this. The exceedingly slow motion of glaciers is well known. A sudden fall of polar snows would immediately and rapidly send its chilling influence over an Eden world; but the full effect of the same would be gradual and depend entirely upon the progress of the glaciers and the volume of snows composing them. Untold centuries might intervene before adjacent lands would yield to the scepter of eternal winter. As the polar glaciers urged their way from the dead to the living world, it changed the climates of genial lands to the lifeless scenes of the glacial epochs.

More than eight hundred years ago Greenland was not the frigid land it now is. Eight centuries ago the Icelanders and Northmen sailed through northern seas, in the interest of commerce, where now our hardiest seamen, in ample vessels, well manned and equipped, scarcely dare to venture. They planted colonies on Greenland's shores, whose very name bespeaks a fruitful clime. They erected monuments on an island in Baffin's Bay whose remains tell a tale of enterprise and energy. They entered Lancaster Sound and Barrows Strait.* Icelandic annals show that their people not only pushed forward commercial enterprises into these now inhospitable lands, but they also carried their religion into the new colonies. Greenland and Spitzbergen were, according to their histories, for centuries prosperous and happy settlements. We must give these annals due credence. What has become of these colonies? Would any nation now attempt to colonize those dreary solitudes of eternal winter, with the prospect of making prosperous settlements? Is it not evi-

---

* See Am. Cyc.: "Arctic Discoveries."

dent that the Greenland and Spitzbergen of the Northmen age were not the Greenland and Spitzbergen of to-day? It really seems that the northern glacier has progressed so far southward that once habitable lands have become desolate—we might almost say "without inhabitant."

It seems likely, then, in view of a former genial temperature in northern lands, that the present glaciation of the polar worlds is but a legitimate result of the decline of the last remnant of outer vapors. From this it necessarily follows that the great ice caps of the polar regions are moving toward the equator, and consequently are continually diminishing. They are continually sending off great icebergs out into the seas, where they melt and drop their load of mud and dirt, gravel and boulders, and, it may be, their entombed and mummied dead.

Thus, it is possible, we are approaching a day when the last iceberg will be borne toward the tropics, and the last glacier be made to loose its grip upon the land, and a more genial clime pervade a greater part of the earth.

NOTE II.

THE LOST CONTINENT.

Evidence is continually accumulating which goes to show that a great Pacific continent now lies under water. Since the chapter on "Oceanic Augmentation" was put in manuscript form I have read, with intense interest, the "Lost Atlantis," by Ignatius Donnelly. The mass of evidence this author brings forth to prove the existence of a submerged continent in the Atlantic waters is simply astonishing. The fact that a great insular continent has, in very recent geologic time, been overflowed, seems to be so clearly proven that it may be looked upon as an established truth. But tell me, how could such great continents sink without drawing the oceanic waters away from the shores of the continents, and thus increasing the pitch of rivers near their outlets? In addition to the submerged continent of the Pacific and that of the Atlantic, between the United States and Africa, there is a vast submerged continent or barrier of the North Atlantic.

These facts leave no room for doubt that the oceans have been augmented, and stand to-day many fathoms deeper than they did, perhaps, in the Edenic day.

NOTE III.

ANTHRACITES IN BRITISH AMERICA.

My readers will remember the claim made in the chapter on "Anthracites," that such heavy forms of coal must lie in great beds along the slopes of British America. In a lecture delivered by the author, before these pages were in the publishers' hands, this language was used: "Long ago I predicted that great beds of anthracite and bituminous coal would be found to underlie the great basin and plateau of British America. England, in the possession of that vast territory, is richer than if she owned all the gold mines of the world."

It is now authentically announced that "a seam of fine quality of anthracite has been found eight hundred miles west of Winnipeg, ou the Canadian Pacific Railroad. The seam is fourteen feet thick!"

NOTE IV.

A SIGNIFICANT ADMISSION.

As these pages are undergoing their last revision before being placed in the printers' hands, a friendly letter comes from Prof. N. H. Winchell, of the Minneapolis University, and State Geologist of Minnesota. In this letter the writer makes the frank concession that the primeval vapors, mineral laden and revolving on high, "must have lingered in the skies much later than has been admitted." This was in reply to my pamphlet, "Alaska, Land of the Nugget, Why?"

It is not likely that the writer ever imagined what a sweeping admission he made. If we admit that the primordial waters did not all come back to the earth till more recent times, then the annular theory of world making is conceded. We admit the geologic ages were outlined by the progressive wreck of the earth's annular system. Then how can we avoid the solution of the glacial problem, for as surely as the vapors fell in modern geologic times they fell as snows in the polar world, north and south. So, too, as surely as those primeval vapors arose from a molten earth they were laden with gold vapors, which readily associate with heated waters, and we are forced to the conclusion that, as these vapors arose together, they rode together for ages in the lofty skies, and, as the earth cooled, they crystalized into forms—snow-flakes, hail, nuggets and gold-flakes, etc., and when they fell they fell together in polar lands. So that the birth-place of the glacier is the birth-place of placer gold.

Primeval vapors, lingering in the skies until recent geologic times, mean an annular system with all that it implies. Then how are we to face the coal and the oil problems—for vapors could not rise from a molten earth and not carry a world of unconsumed fuel with them? It is of the utmost importance that geologists be careful how and what they concede. The earth in all its features is linked to an ancient ring system. Go what road we may, we are perpetually meeting with its deathless survivals. The miners of Alaskan gold all admit, as they collect the nuggets and the dust from the frozen earth, from the very summits of mountain peaks and from the surface of glaciers, that the precious hoard was not ground from the rocks. Whence came it then? A letter lying before me affirms that the "gold found on the surface of Mt. Fairweather Glacier must have come from the heavens." How the truth struggles to the light!

THE TRUE ORIGIN OF COAL—THE VEGETATION THEORY DISPROVED.

The following lecture was delivered before the Belmont County Teachers' Institute, at its meeting in Barnesville, Ohio, August 11th, 1885:

I am glad to meet with you, my friends, teachers of my native county, in the discussion of the coal problem. I am glad because it is to the student and instructor that I must confide the promulgation of the theory I am about to explain. It is a most important problem; important, because it involves the fate of many time-honored theories. This is a day when all scientific questions are tested by the calcium light of reason—weighed in the philosopher's scales, and valued by the microscopic test of law.

A theory that absolutely fails in one point is a complete failure. And, after a critical examination of the hypothesis that the coal beds of the earth are a vegetable product, it is found that it does absolutely and utterly fail in many points, and that it has scarcely a feature that can abide these tests.

The Chart.—I present before you a chart representing a planet just issuing from the igneous or molten condition—a planet constituted as the earth, of water, mineral and metallic matter—a chart exhibiting the great and essential fact that such a planet must at some time in the course of its evolution become surrounded by a complex system of equatorial rings, composed of matter concentrated from its nebulous empire and mingled with aqueous and mineral vapors, expelled from its heated center during the reign of fire. It represents the original atmosphere of a burning world, which atmosphere formed into rings as the suspended vapors condensed. So that such a system is necessarily composed of matter, meteoric and vaporous, on its downward course to a common center, and other matter driven outward and upward by the measureless energy of heat, and which, from utter necessity, upon falling within, becomes no inconsiderable part of the planet's sedimentary beds, or aqueous-formed crust. It presents to your view what I have called the "Annular Theory." —a thing conceived in my boyish mind as I gazed in confused wonder upon the ring system of the planet Saturn. Believing that one unchanging and universal law presided in the construction of worlds, I could not divest my mind of the conception that the earth also must have had at one time a ring or annular system, and amid the "ups and downs" of a varied life the idea has lived as a part of my mental being, until it now seems to be a positive reality. In the course of time the question of the origin of coal, with numerous other important ones, became involved in and, as I hold, satisfactorily explained by this theory, after all these crucial tests had been applied.

Now you will excuse me if I should be a little tedious in approaching this question, for it is no ordinary one—one upon which you can afford to devote, not minutes nor hours, but days and weeks of thought.

The first thing for us to understand, in the evolution of this theory, is the now settled fact that this earth was once a burning orb. Wherever we turn the telescope upon the universe of worlds we see the glowing suns of a scintillating creation—the sparkling centers of evolving worlds. The spectroscope also speaks, and in it we put implicit faith, for it cannot falsify. It photographs on the philosopher's screen the flames of comets and suns. It tells us that nebulæ, planets and stars are composed of materials the same in kind as those out of which the earth was built. Every star is a burning world, and consequently a smoking world—a mighty crucible in which mineral and metallic elements are fused, vaporized and sublimed; where chemical compounds are reduced and re-combined; where the work of creation and re-creation is going on forever.

Now let us imagine a world composed exclusively of water and sandstone. Let it be fused or melted to its inmost depths by inveterate heat, as millions of worlds are to-day. By this heat its waters would be vaporized and driven away from the fiery mass, and the core would be a mass of melted silica, and an atmosphere of aqueous vapors would surround it. Now, if the heat be increased so as to make the mass a shining sun or beaming star, the silica would be vaporized and also driven away and made to commingle with the watery vapors, and, if the melted mass contained limestone, iron or lead, these substances would also be vaporized and the vaporous atmosphere would contain gaseous matter of all these substances. And it must be seen that in the universe of law, as the mass becomes cold, these vaporized elements would condense, in order of their susceptibility of fusion and vaporization, and, falling to a common center, would form a spherical mass, not of water and sandstone as before, but one composed of all these elements. It must also be seen that the aqueous vapors would be the last to condense, and, moreover, the last to fall from the position which they must have taken under the reign of heat, repelling from a focus; and, while all these materials thus vaporized and afterwards condensed, must to some extent become commingled and form, just as we see under our feet, a heterogeneous world, yet there would be upon the whole some definite and regular order of strata arrangement. For, in the condensation and consequent precipitation the heaviest and most refractory minerals—minerals most difficult to fuse and vaporize—would separate from the rest and settle first. Beds of silex, almost pure, and silicious beds containing iron, calcium and every other metal or mineral contained in the fiery envelope would recur in some kind of order in our hypothetic world. There would be beds of metals arising from this fiery distillation nearly pure, under the law of elemental assortment and segregation. We see this law abundantly and universally exemplified in the entire structure of the earth's crust. We have sand beds, lime beds and metallic beds, all nearly pure; and then beds of every degree almost of impurity or mixture. Hence, we are at once forced to face the self-evident truth that there must be in the sedimentary beds of every previously molten orb the very material that formerly had existed in its atmosphere. If that atmosphere or glowing envelope contained carbon in any form whatever, that carbon would be contained as pure and also as mixed beds in the sedimentary crust. Now let me ask you to remember this vital fact. I will state it again, so that you can all certainly understand it. In every world whose fiery and vaporous atmosphere contained carbonaceous matter there must be beds of carbon of varying degrees of purity. For the same reason that silicious minerals must form sand beds as they fall from the great vaporous fund; for the same reason that calcareous minerals must form strata of limestone, mineral carbon, in whatever form it existed, whether as graphite or carbonate, as carbonic anhydride, or even as the diamond, must have separated in that inevitable sublimation from its associated elements and finally formed carbon beds in the evolving world. Beds of primitive carbon! Carbon that needed not the light of the sun nor the mysterious laboratory of the plant to make it carbon. Carbon that existed as such millions of years before a plant cell existed.

Did I call this a vital question? Is there a man who dares dispute it? It is vital because it utterly abrogates the old idea that a carbon bed can have no other source than that of vegetation. Here is the critical foundation upon which geologists stand to-day, who say coal is a vegetable product because vegetation is the only competent source of carbon beds. Away down amid the archæan piles of the earth, amid primitive rocks that never felt the thrill of the sunbeam's touch; where never a plant, a twig, a leaf, or a bud can be found as a fossil, you will find stupendous beds of primitive carbon, as all geologists know full well. Then, I say, please remember the vital fact, while I go around it and approach it from another side.

A molten or a burning world, rotating upon an axis, as the earth does, will fling its great atmosphere of vaporized water, mineral and metallic matter into rings in its equatorial regions. I cannot, in the short time I have to address you, give very much evidence to prove this proposition. It is, however, susceptible of the clearest mathematical demonstration. I will simply state a few facts and ask you to admit it, and, if you are ever fortunate or unfortunate enough to read the "Earth's Annular Theory," you will find the startling demonstration there in full and so simple a child can understand it.

The exhaustless and measureless energy of heat exerted to vaporize the refractory minerals and metals, now glowing and sparkling in millions of stars—grand central fires of other systems—must drive these vapors so far into space that, as they necessarily obey the mighty impetus given them by the orb's rotation, they accumulate so much energy that they must continue to revolve independently about the central body for a long time after the mass becomes cold. The day is not very far distant when it will be admitted on all hands that ring or annular formation is an indispensable part of planetary evolution.

We see this necessary and legitimate result of plutonic energy beautifully and grandly exemplified in the clockwork of the skies. Jupiter and Saturn, twin giants of the solar system, proclaim this eternal truth across the mighty void that separates this puny world from them. See how gloriously they thread their course through the heavens, Titanic worlds yet unfinished. A great part of their oceans and much of their future sedimentary crust are yet revolving about them as vapors and meteoric dust. Jupiter has belts, material belts, revolving about him, and Saturn has both rings and belts. It can be readily demonstrated by physical law that Jupiter's belts were once in the form of concentric rings in his equatorial heavens. Also that Saturn's rings are gradually approaching the planet, and that the belts of both planets are gradually falling to their surfaces by way of the polar regions. So that the day will come, as sure as law presides in the government of heaven, that Saturn will be stripped of her glorious appendage.

Did the earth ever possess such an appendage? There is not an astronomer who will say no. There is not a geologist who will not say it did after he shall once have examined the geological record with an impartial and philosophic eye. Examine the world upon which you live! See what stupendous revolutions are chronicled in its rocky volume! What is its past history? A thousand volumes cannot reveal it all. But there is one chapter that I will attempt to interpret to-day—a noble chapter, written with a pen of fire on immortal stone. What does it say? On its title page we read: "When the dial finger of time pointed to the dawn of ages the earth was a burning world." It rolled through space a glowing sun. Its rocky beds were molten and the oceans that now wash the rock-bound shores were held in suspension on high by the repelling power of heat. This igneous or fiery condition of the earth in primeval times is admitted on all hands. If any question has ever been settled by the philosophy of man, this one has. Then the earth was no exception in its mode of evolution in the universe of worlds; and all its present oceans and a part of every substance now found in its upper crust existed in its great primeval atmosphere. What a wondrous atmosphere that was! Twenty miles of aqueous strata tell us what it was composed of. Mixed with a mighty fund of silicon, calcium, iron, copper, lead, silver, gold, sodium, oxygen and hydrogen was an immensity of carbon. All the carbon now in the lime beds of the world's crust was there; all the carbon in the carbonate of iron, zinc and lead, and all the carbonates were there; so also the measureless fund of carbon now stored away as coal was there.

But the more extensive the primeval atmosphere the more likely are its condensing vapors to be whirled into rings. The rim of a wheel must rotate in the same time the rest of it does. And, after making all due allowance for the mobility of the mass of the great fund of vapors that ever canopied the primitive earth, it must be admitted that the boundary of the same moved with immense velocity. Now, many eminent men of science advance the claim that the atmosphere was at least 240,000 miles deep. We will not claim half that depth—say 100,000 miles; but the peripheral boundary of an atmosphere of that depth, even, if the entire mass rotated once in twenty-four hours, as our atmosphere now does with the earth, had a rotary velocity of 25,000 miles per hour. But this is nearly 8,000 miles per hour more rapidly than it need to move in order to whirl the vapors into rings, even if it was not one mile deep or at the surface of the earth.

By means of Kepler's third law it is an easy matter to determine how deep the atmosphere must have been in order that the revolution of the same in twenty-four hours would throw its condensed matter into equatorial rings. Any one who will take the trouble and make the calculation will see that it need not be 240,000 miles, nor 100,000 miles, but a little more than 7,000 miles. Hence, we are forced to the conclusion that all the primeval vapors situated more than 7,360 miles from the earth's surface continued to revolve about their primary center after they grew cold and condensed, and all those situated nearer the earth fell to its surface because they had not centrifugal energy enough to retain them. Such are the demands of law, and I presume there is not a mathematical or philosophical mind that can doubt this conclusion after having sufficiently contemplated the facts that the earth was once in a molten state and rotated as it now does. A neglect or failure on the part of the geologists to follow the effects of this fiery condition of the primitive earth to this legitimate end has led us into pernicious errors. Now where are we? We found an immensity of carbon in the primitive form in the fiery envelope of the earth; and we now find that it existed in the annular form revolving about it, associated with its aqueous oceans of vapor. These two facts stand out prominently in the annular theory and challenge the world for a refutation. You already know that the whole course of geology has been pursued with the idea that the waters and their associated matter all fell to the earth before any of the aqueous beds were deposited; that is, they never formed into rings. Now, if these vapors never formed into rings, then geologists are right, and I am wrong. But a molten world—an igneous era—necessitates ring formation; then, if I am wrong in my conclusions, the earth was never in an igneous condition. But it must be said that we simply know it was once in that condition. There is where we are—simply forced to the conclusion that elementary carbon, once driven from the fire-born earth, revolved about it, and, therefore, fell, with its aqueous vapors, in grand instalments. Now keep these facts in view while we make another excursion and bring up other reinforcements.

Every philosophic mind will agree with me in the claim that if the earth ever was in a burning state, it was also in a smoking state. The constitution of the globe is such as to render this a necessary and absolute fact. From every fire-place and furnace, from every volcano on earth, issues smoke. But smoke is unconsumed carbon. This carbon is in the form of infinitesimal particles, and, being released from various combinations, is, according to chemical law, in what is called its nascent state and eager to enter into new combinations. Hence its ready union with the oxygen of the air, as any chemist can prove, by which smoke becomes invisible in a short time. But nascent carbon or smoke has an affinity for hydrogen, and if it cannot obtain oxygen it will dissolve or decompose aqueous vapor and appropriate its hydrogen, forming a hydro-carbon, after which the oxygen just released combines with it, forming an oxy-hydro-carbon. Were it not, then, for the presence of oxygen in the air, the mighty volumes of smoke that eternally arise from millions of chimneys would, in a short time, fill the atmosphere with midnight blackness. There would be a constant deposit of hydro-carbon in the form of soot upon the earth, and men might collect this carbon and burn it, as we now do our coal.

Let any one watch the black column of unconsumed carbon issuing from a locomotive burning bituminous coal. In how short a time it vanishes in air. It is simply undergoing a second combustion—a union with oxygen. Or you may perform a more simple experiment by burning a rag or piece of paper in the open air.

The carbonized paper or rag lying before you may be lighted and made to burn again. A slow combustion will spread over it again and again, until the carbon becomes invisible, leaving nothing but ashes. You have seen this secondary combustion many a time in your soot-clogged chimneys and stove-pipes. You have seen a transient flame play over the back wall covered with an oily soot.

These are manifestations of unchanging law. They exhibit an ocular demonstration of the fact that smoke or unconsumed carbon, from whatever source it comes, is again formed into fuel, and will burn again. And if, as it arises from the furnaces of the earth, it could be stored away among watery vapors, where oxygen could not play upon it, it would all be burned into an oxy-hydro-carbon. What is an oxy-hydro-carbon? It is an oily or bituminous substance, composed of carbon, oxygen and hydrogen—the very substance stored away as coal in the earth's crust. We are now ready for another vital question. What has become of the vast fund of smoke that went up from the burning or igneous world? What has become of the unconsumed carbon distilled by the earth's inveterate fires? Every man must know that it went up and lodged, so to speak, among the suspended vapors on high, and, being in immediate contact with them, in a very ocean of hydrogen, it must have become a hydrocarbon. Now it does not make a particle of difference whether the smoke issues from the flying locomotive, from Ætna's fiery entrails, or from millions of telluric flames, the distillation of carbon is the same, and it can make no difference whether it hovers in the atmosphere or is flying with volcanic force thousands of miles into space; it is governed by the same immutable law. In the atmosphere the oxygen devours it, and it vanishes; but beyond the atmosphere, under the cope of heaven, among the whirling vapors, it lodges as soot-black hydro-carbon.

Now, where are we? First, we had carbon as a primitive element, before the dawn of vegetation. Therefore it was not a vegetable product. Next, we find it revolving in the earth's annular system as carbonaceous rings, and vegetation did not put it there. Next, we find those rings to consist of a black, sooty, oily, pitchy hydro-carbon, sent up from the fiery focus of the planet, and yet vegetation has taken no part in the grand metaphysis. Keep these facts in mind awhile.

The hydro-carbon seen by the mind's eye as dark rings and bands surrounding the primitive earth, just as the dark carbonaceous bands of Jupiter and Saturn surround those primaries to-day, and the hydro-carbon buried in our rock-ribbed hills, are one and the same thing. The vapors have fallen, and the carbon must have fallen with it. Titan hands have gathered it and stored it away for the use of man—he mines it and burns it again.

Now, why should men conclude that vegetation is necessary for the formation of carbon beds, when this element, it must be admitted, existed as a combustible fuel before a plant germ ever existed in the earth? The formation of carbon from peat moss is a combustion or redistillation precisely similar, except in degree, to that which took place amid the aqueous vapors on high. Could we take the carbonized paper or rag in our experiment before it burns again, and place it side by side with peat-formed carbon, they would be precisely the same in kind. As the peat moss decays—or, in other words, is consumed—a charred product, which falls amid the waters of the bog, remains unconsumed, whereas, if it had remained in the open air, it would have vanished in air. So that the slow combustion in a peat bog becomes our first witness; and a very important witness it is. Its testimony is that, if the puny combustion that takes place in the decay of peat moss can produce carbon in small quantities, just as soot is formed in the combustion of wood or any other carbonaceous substance, then the mighty, stupendous and Titanic combustion of archæan times must have produced an infinite and measureless amount of it, and, being a primitive distillation, it must have made a purer product.

It must be with a full knowledge of the fact that peat carbon is but a secondary transformation that the vegetarian takes this primitive carbon, already a combustible fuel, and made so by inexorable law, but disregarding the smoking furnace of the infant earth, conceives it to be the unburnt product of peat moss, when he must know that so sure as this earth was once in a molten state its rising carbon vapors were changed into fuel carbon among the aqueous vapors on high. He simply substitutes for that grand distillation the slow decay of vegetation. Now every one in the house must see that a necessarily stupendous production of fuel carbon is thus abnegated by a mere triviality. Must we deny the testimony of the geologic record; the testimony of our sister planets; the evidence of the sun and stars, in order that the vegetarian may conceive that to a peculiar class of plants are delegated these grand offices of world making? You see that if he admits the agency of telluric heat, he already has carbon fuel on hand; and if he calls in the aid of the spontaneous fires of the peat marsh, he only substitutes an utterly inadequate process to make what was made before.

Now I must give a little history. During a long and tedious series of experiments I demonstrated that the soot arising from burning wood or coal would dissolve aqueous vapor and become a fuel. Then I knew that the smoke that arose from the burning world and entered the suspended vapors did the same. I had a little sack of soot stored away in my laboratory and cabinet with which I was experimenting. While I was eager to satisfy myself that soot becomes a hydro-carbon in the air, nature was secretly at work in an effort to draw this work to a conclusion, for the sack of soot had become so far hydrogenated that it one day began to oxidize in earnest. It took fire spontaneously, thus proving the very thing I had been claiming. It was a dearly-bought demonstration, but it was conclusive. In that building I had placed my geological cabinet; specimens, many hundreds of them, more valuable to me than gold, the work of thirty years of search; my telescope, the work of my own hands —all my tools and drawings—ruined or destroyed. It seemed a little severe that all these things should be offered up, a burning sacrifice, to prove that the unconsumed carbon arising from every fire place, from every volcano, from every planet, star or sun in the universe, became a fuel hydro-carbon. As I looked down upon the ruins a voice seemed to whisper, "It will be beauty for ashes."

That conflagration proved to me, and proves to all men, that somewhere in the earth's crust must exist a combustible form of carbon, sent up from the igneous world. Where is it? Where is it? The man who answers this question solves a momentous problem.

GRAPHITE.—It is evident that if coal be a vegetable product, all other carbon beds must also be of vegetable origin, so that geologists agree with Dana that those vast beds of graphite, found in the oldest aqueous formations, are as much a vegetable formation as peat itself. But right here they meet a stumbling block which they can neither leap over nor circumvent. The laurentian graphite was deposited in an age when vegetation, so far as can be determined, did not grow. No vegetable fossils can be found either in the graphite itself or its associated beds. Geologists have never found a reliable or satisfactory trace of a plant in ancient graphite or near it.

Now it will avail nothing to advance the claim that vegetable fossil prints have been obliterated from the crystalline rocks, for in these very rocks the delicate form of the eozoon has been preserved.

Considering the immense amount of archæan graphite, it is simply impossible that some traces of plants should not be found in it or near it, if it were a vegetable product. Let us examine this form of carbon in the light of the annular theory.

Whenever carbon is distilled, either from wood, oil or bones or limestone, many allotropic forms of carbon may be obtained. We call them light and heavy forms. These are interestingly illustrated in the manufacture of burning gas from coal or coal oil, or in the process of refining crude petroleum. Here heavy forms of carbon are distilled as asphaltum graphite, so that it is well known that these are necessary products of all such distillations. Hence the igneous world must have yielded all such allotropic carbon, and the heaviest forms, as graphite, etc., must have existed almost exclusively in that part of the annular system nearest the earth, and must, therefore, have fallen and became incorporated among the oldest sedimentary rocks. Its weight must have located it near the earth, according to law, and its position must have necessitated its fall to the earth before the lighter forms. Its present position is simply a result of annular arrangement previously determined by law in the vaporous atmosphere. Thus, you see, while the old theory utterly fails to account both for the origin and position of graphite, the new one shows why there are no traces of vegetation there, why it occupies the position it does, and gives the only philosophic origin that can be given. Now, suppose it was not found in these early-formed beds, but bodily among the carbon beds of the tertiary rocks. Any one can see that the annular theory would fail to account for it. But primitive graphite can never be found in any other position, except in small quantities. Hence, we must admit its intimate connection with the annular system. Thus, in support of the annular theory, I find the graphite just *as* I want to find it, and, moreover, just *where* I want to. The vegetarian finds it as he don't want to, and where he don't want to. Choose ye this day whom ye will believe.

An Aqueous Formation.—The general appearance of a coal seam is that of an aqueous deposit. Examine it where you choose—among the Appalachian metamorphic centers or at any point in the undisturbed beds of the Mississippi Valley, pocketed in prairie lands or stored in the everlasting hills, or in that grandest and greatest of coal fields stretching from Mexico to the north polar sea—you cannot close your eyes to the universal evidence that all these coal seams were planted at the bottom of the sea, and not in a peat swamp at its surface level.

It is everywhere planted immediately upon a sea-formed bed. Thin, aqueous seams or partings run through it, and a sea-formed bed is planted upon it. And, if the coal in the very center of these deposits is a vegetable product, there is no human eye that can tell where the sea deposit ends and swamp formation begins. Every day, almost, I meet with these evidences. Sometimes a limestone stratum is almost the immediate covering of a coal seam, and again almost its immediate base. But limestone formations were deposited in the quiet seas. There are in this part of Ohio more than a dozen coal seams, and extending from here to the Ohio River, Belmont County is underlain with a dozen heavy lime-rock strata, and these are interspersed among the coal beds; and, more frequently than otherwise, they are in close proximity to the coal. Vegetarians say these things require that the "waters of the ocean should be very near." This forced admission is more amusing than philosophic, and yet I must agree to it. The waters of the ocean were very near—very.

Boulders.—Again and again boulders are found embedded in the coal in such position as to show that they were carried to the spot in floating ice, or trees, at the very time the bed was forming. This, as any one can see, demands that the bed was forming at the bottom of the sea. In the museum at Columbus is a boulder found in the midst of a coal seam. It was torn from its native crystalline bed, worn and glaciated by ice, before it was imbedded in the coal. Our State geologist says, "Ice transport would seem almost necessary for such a block," and all agree that the waters in which it was transported "were very near."

Why should it be "almost necessary" for ice to be involved in the transportation of ice-worn and ice-borne boulders? Simply because a fuller concession involves the reputed origin of coal in irretrievable contradiction.

Coal Partings.—I would I had an hour to devote to the consideration of the thin, laminated partings in coal, sometimes not thicker than paper, frequently not the eighth of an inch, and sometimes half an inch, extending over thousands of square miles. There is not a living man who can account for the mysterious features of these partings by the current theory without opposing law and reason. If, for instance, the lower bench of a divided seam is a product of vegetation, the roots of which were planted in the clay bed below, where is the root bed or the upper bench? In a clay parting of half an inch thickness? Where is the vegetation that this thin clay or sand parting ought to have involved, as it accumulated over the swamp plant, etc.? Blades and twigs and moss stems should rise vertically through such partings, for the mud and sand must have been deposited around them. But the vegetation was not there! There can be but one conclusion: These partings would show the vegetation if it had been there.

NUMEROUS SEAMS.—We have in Eastern Ohio more than a dozen coal seams, varying from a few inches to nine or twelve feet in thickness. A twelve-foot seam required nearly one hundred feet of vegetation—so said the illustrious Dana. In Nova Scotia are more than seventy different seams; one of them is thirty-five feet thick, another fifteen, and another twelve. Two hundred and forty feet of accidental soot or unconsumed carbon, according to the current theory, escaped the peat bed combination and formed Nova Scotia's heaviest bed of coal. More than one hundred seams are found in England. According to the current theory, each bed records a submergence of a peat swamp, and subsequent re-elevation—a re-elevation just above the ocean's level, seventy or one hundred times, is a remarkably accommodating series of telluric changes.

Now, let us take another glance at the annular system. You see it, darkened by carbonaceous rings, bands or belts, every one of which must fall to the earth as a great installment of carbon, and each must float away and be deposited in the sea. If it be a true representation, it first corrects the impression that the formation of carbon beds belongs exclusively to any age. While certain periods or ages are characterized by a more abundant downfall than others, it is plain that, as smoke and vapors of water are so nearly of the same gravity, all aqueous vapor of the igneous age must have more or less been mixed with carbonaceous matter. The geological record is positive in the declaration that carbon was more or less deposited in all geologic times, from the very day in which the heaviest form of graphite fell and was buried in the heavy metalliferous beds of archæan time to that grand deluge of modern time, yet living in the traditions of man. We can, therefore, no longer marvel at the number of carbon or coal veins. As superaerial carbon and other matter must, under law, fall largely in the polar regions, we can no longer wonder why the number of coal beds increase toward the North; no longer wonder that heavy deposits of coal were made in lands now locked up in eternal ice.

EQUATORIAL COALS.—If coal be a vegetable product, then surely in those regions of the earth where vegetation in all time has been the most luxuriant we should find the greatest development of coal. Then, if we find such vast beds as I have before alluded to, in the latitude of Nova Scotia, in the valley of the Mackenzie, in the hill lands of Siberia, what must we expect to find in tropical lands, under the equator? But where is the coal of the tropics? It is rather scarce. Now, can it be possible that there has never been an opportunity offered, in the perpetual oscillation of sea and land, that the vegetarian claims for swamp and peat formation in the tropics? Have the equatorial lands during these subsidences and elevations always been too deeply covered with water, or elevated so far above the waves as not to permit swamp vegetation and coal formation? Why were all these wondrously-accommodating changes of sea level confined to colder regions? What a happy circumstance it would be if the vegetarian could point to such great coal basins in the tropics as we find almost everywhere beyond them! Now, if he could do this, it would be a bad thing for the annular theory. I would be forced to admit that it essentially failed in one particular, and was therefore a failure. But here, as in the case of the graphite, the coal is nearly absent from the very lands which the new theory claims must contain the least. Tellurio-cosmic matter could have fallen directly in very limited quantities in the equatorial regions, more largely in the temperate zones, and in the greatest quantities in the frigid zone. This is susceptible of the clearest mathematical and philosophic demonstration. All the coal that lies within the tropics must have been borne thither as carbon dust from other lands or other waters, and all the coal in these lands must necessarily have the least specific gravity. Those coals in the southern part of the north temperate zone will in turn be found specifically lighter after a fair test (and an elimination of foreign matter) than those in the northern part, and the nearer we approach the polar world the heavier and purer we will find the coal. I will give the theory over to these final and decisive tests. If the atmosphere were filled with cosmic dust to-day this matter would fall most rapidly at the poles, because gravity there is the strongest. In addition to this, the centrifugal force in the rotating mass at the equator would somewhat check its fall there. The upward currents of air, also, in the torrid, and the downward motion in the temperate and frigid zones are distinct causes operating to decrease the fall under the equator and increase that at the poles. But terrestrial belts, like those of the Jovine and Saturnian worlds, would, according to law, move from the equator to the poles before they came down into the lower air. For this very reason, then, the annular theory would utterly fail if coal were found more abundantly in lower latitudes than the middle temperate regions. It would fail if we did not find it in abundance in extreme polar lands. It would fail if the northern coals were not heavier and purer in carbon. Now the vegetation theory utterly fails to explain these anomalies, and its advocates will grope in darkness and error as long as they cling to it. Long ago I predicted that great beds of anthracite and bituminous coal would be found in coming ages to underlie the great basin and plateau of British America. England, in the possession of that vast territory, is richer than if she owned all the gold mines in the world. These are only some of the decisive tests of the correctness of this theory. Their name is legion. I will select one more and close this evidence.

CARBON IN ICE.—The aqueous vapors above the firmament must have fallen in the polar world as terrific downfalls of snow. The black carbonaceous matter of the system must have fallen with them. Such a downfall to-day in the polar regions would soon dissipate the great polar ice caps, on account of their absorbing solar heat, and the extreme arctic and antarctic climate would be so ameliorated that vegetation would cover all their land areas as it did the great northern continent in pre-glacial times. A slight downfall, however, would not affect climate to any great extent. The northern snows that formed into glacier ice would also contain the carbon dust. But if we should find great carbon beds frozen up in everlasting snow and ice, is there any man upon this planet who would claim that it was a vegetable product? But these ice-imprisoned beds are simply facts of ocular demonstration. To-day carbon beds may be found along the shores of the north polar seas. On the coast of Kotzebue Sound rise masses of ancient glacier ice above the waves. This ice is planted on seams of carbon, and, what is more astonishing, just beneath the carbon beds are solid beds of ice, extending downward as far as the ocean waves have permitted examination. Now what does this mean? We might find carbon beds in rocks baked and crystalline under the reign of fire, and imagine it the product of vegetation. But here, in the snow-bound islands of the polar deep, packed in eternal ice, ice above and ice below, we will bury the old theory and erect the new. There it is, just as the annular theory demands. Carbon beds found planted upon a pavement of ice, just where the new theory wants to find it, and just where the old does not want to find it.

A vast fund of carbon is found deeply buried in the frozen mud and sand in Northern Siberia. At Yakutsk, and in other places, deep wells have been dug or bored, and the deepest have never reached the limit of eternal frost. The strata, so far as the auger has explored, are alternating beds of frozen sand, frozen mud and frozen carbon, of course called peat. It is said that this part of Siberia is solidly frozen to a depth of at least six hundred feet. Can it be possible that the earth could have frozen to that depth after the carbon beds grew as peat vegetation? It cannot be. These beds must have frozen as they were built up. Did peat grow in that way? Thus, in every nook and corner of the earth, we find the strongest evidence of the truth that the various forms of carbon now found in the crust of the earth is almost wholly a primitive product of igneous time. But let us remember that the coal question is only a small part of the grand problem. The annular theory had been proven to be true before the coal was ever thought of in connection with it.

A little reflection will now lead you to the conclusion that the only difference between the geologists of to-day and myself is that they pull one way; I pull the other. We all started in the same direction. I am going the same way that I started. We parted company on the confines of paleozoic time. At that point they said all the matter in the earth's primeval atmosphere fell previous to that time. I said but a small part of it had then fallen, and the rest remained revolving, as the Saturnian rings are to-day. The difference between us in the beginning seemed slight, indeed. How easy it would have been to have admitted the truth then! Had it been done, almost every department of physical science would be to-day a century further on its way. But, slight as the difference was in the beginning, there has been a great divergence in the lines to important conclusions. One has led to error and falsehood; the other to immaculate truth.

The old school utterly fails to explain some principal questions in geology; the new challenges the presentation of a geologic question it cannot explain. The former antagonizes law; the latter demonstrates and defends it.

Now, if need be, I will set the carbon question aside and prove the truth of this theory from other geological evidence without its aid. Nay, the whole field of geological evidence may be disregarded, and the theory proven by astronomical science, backed by mathematical law. Or, if you choose, we will set aside all scientific evidence, and the first eight chapters of Genesis will champion the cause and prove its truth, even to a world of Voltaires, Paines and Ingersolls. Oh, the grandeur of this amazing field of thought! I invite you to enter this untrodden region. If we do not explore it, let me assure you, other men will do it in other times. I have put my hand to the work: I cannot turn back, and shall welcome all men to my assistance. I have already detained you too long, but I knew it was to the teachers of these times that I must expect to delegate the charge and championship of this theory, that must meet the marshaled hosts of veterans that will oppose it. The young and unfettered intellects of this and coming days will see the ruins of the old theory. But the change will be slow. Twenty-five years ago, when I began to advocate the annular theory, men smiled over the attempt of a young man to claim the discovery of anything new in geologic or astronomic science; and when their attention was directed to the annular system of Saturn as the grand key that unlocked the deep mysteries of planetary evolution, they said there was no force in the claim. When told that the primitive earth, molten and beaming as a sun, necessitated ring formation, and that astronomers and geologists, in their reckoning, had never taken into account the potential energy stored up in the annular vapors revolving on high, in that day of fervent heat, they still remained silent. When asked how the oceans returned to the earth, after they hung for unknown time thousands of miles beyond the atmosphere of the day, still no answer came. When told that the oceans, in great part, fell to the earth in deluges of terrific violence in modern geologic time (sub-silurian), the idea was visionary and anti-scriptural. When told that the coal deposits of the world could be nothing more nor less, under the demands of law, than aqueous deposits of primitive carbon distilled in Vulcan's mighty alembic, and which was sent up amid the aqueous vapors where it became a hydrocarbon, thoughtful men hegan to think; others turned away to laugh.

When scientists put forth the astounding information that a mighty ice sheet, thousands of feet thick, pushed down from the far north and leveled our hills and filled our valleys, obliterating rivers and lakes, and they were told that such a fund of snow must have fallen from telluric rings of vapor in the polar regions, some men said a profound mystery has been solved; others said, "Oh, no." Finally, when, in 1874, I published a little volume presenting some of these questions more generally to the public eye, and especially called attention to the fact that the great deluge of Noah was a down-rush of the last remnant of annular waters, I received some of the most flattering commendations on the one hand, and on the other some of the most excoriating and denunciatory criticisms for thus "treading on holy ground."

Now what is the result of not admitting the unavoidable conclusion that the oceans did not and could not all fall in primeval times? I need not tell you. You see the world is hard to be moved, but it does move none the less, and let me tell you, though the thought may savor of vanity, it is moving into a new orbit. It is ripening—ripening for a golden age of thought. It is drifting to the border-land of the annular theory. How could it be otherwise? Law, eternal and inexorable law, is its foundation, and thither it must gravitate.

In conclusion, then, let me say I have passed over this amazing field of thought so often, and have found so much incontrovertible evidence that I no longer care for nor fear opposition, come from what source it may.

This world was once surrounded by an annular system. The coal now sleeping in the crust of the earth was a part of that system. See the amazing wisdom of God in thus gathering it from the fiery center of the earth for man, storing it away near the surface, where alone he can get at it. Had the earth never been in an igneous state, all the carbon must have been disseminated through the entire earth, and there could not have been a true coal bed. When men will open their eyes and look they must see that the burning, smoking, boiling and seething earth must have made measureless quantities of unconsumed carbon, and they must see that they have been clinging to and preaching a pernicious error. Geology will then be placed upon its true and immutable foundation. Genesis will then stand the test of philosophy in the view of all men. Revealed and natural religion will walk hand in hand, and even the skeptic must see that the first eight chapters of Genesis contain in simple but unmistakable terms a complete and positive demonstration of the annular theory, thus conveying in the noblest thoughts some of the grandest truths, revealing an Edenic world the most fascinating, and a sunlight the purest that ever illumined the world. Oh, haste that glad time! Haste the day and the hour when the unnatural conflict between the theologian and scientist will cease forever; when the discordant elements may be allowed to rest in the calm repose of death; when man, fallen and weak, may arise more nearly in the image of God, the great Philosopher and Architect of the material universe, whose mighty hand has ruled the evolving world in all time, whether swaddled in flames and rocked in a cradle of fire, or blooming in the pure sunlight of eternal day.

CAPTAIN CARTER'S ORIGINAL DEMONSTRATION.

On page 27 of this volume is a simple calculation, showing the distance of a terrestrial ring from the earth's surface. Prof. Carter took up the same calculation and obtained the same result within a very few decimals, and with this calculation sent the following demonstration by a new and original method. These calculations prove beyond the shadow of a doubt that vapors in the primitive atmosphere, at the distance of about 26,000 miles from the earth's center, had a velocity in revolution that kept them there, while those vapors nearer the earth fell, and hence the earth had a ring or rings. Thus it is susceptible of the clearest mathematical and philosophical demonstration that this earth was once surrounded by a system of Saturn-like rings and belts; that the aqueous-formed crust was to no small extent built up under its influence cannot admit of a reasonable doubt; and the measureless periods of geologic time are thus necessarily shortened.

PROBLEM. Required the orbital radius or distance of a satellite of the earth, the time of revolution being given.

Let $f =$ centrifugal force of satellite (ring), $v =$ velocity of same, $n =$ uumber of seconds in time of revolution. $R =$ mean orbital radius of satellite, or distance from the earth's center, $g =$ gravity at earth's equator, $g' =$ gravity at satellite or ring, and $P =$ earth's equatorial radius.

Now $V = \frac{2\pi R}{n}$ from Mechanic's formula (1).

Also $f = \frac{V^2}{R}$ from Mechanic's formula (2).

Likewise $g' = \frac{g P^2}{R^2}$ from Mechanic's formula (3).

Substituting (1) in (2) we have

$$f = \frac{4\pi^2 R^2}{n^2 R} = \frac{4\pi^2 R}{n^2}$$

But $f = g'$ from the conditions of the problem,

$$\text{hence, } \frac{4\,\pi^2\,R}{n^2} = \frac{g\,P^2}{R^2} \text{ or } 4\,\pi^2\,R^3 = g\,n^2\,P^2$$

$$\text{And } R = \sqrt[3]{\frac{g\,n^2\,P^2}{4\,\pi^2}}$$

Now, in this case, $P = 20{,}923{,}600$ ft., $n = 86{,}164$ seconds, $g = 32.1937$ and $\pi = 3.141592$, and therefore,

$$R = \sqrt[3]{\frac{32.1937\,(86{,}164)^2\,(20{,}923{,}600)^2}{4(3.141592)^2}} = 6.61$$

times the earth's equatorial radius, $= 26{,}194.01$ miles.

R. K. CARTER,
Chester, Pa.

THE ORIGIN OF PETROLEUM.

A LECTURE, BY PROF. I. N. VAIL.

Los Angeles, Cal., June 21st, 1900.

In this hall, not long ago, I listened to a learned discourse by a prominent geologist of the current school on the origin of the oily carbons. We were told that the immeasurable ocean of oil now locked up in the rocky bosom of Mother Earth "was certainly an organic product, and that the organisms involved in its formation were principally fishes." To this I may add: Such is the prevailing opinion of geologists to-day, and, in view of the high regard I have for many workers in this vast field, I am not disposed to molest them in their blissful attitude, though I want to assure my audience here to-night that I am not present to talk on fish oil, but to give some new and burning testimony, which more than thirty years of deep study of nature's eternal processes have linked to my very being. We have come to hear a lecture on the "Origin of Petroleum," or rock oil, a mineral product of the mineral earth—a product with which, as I see it, never a fish nor a mollusk, nor any other organism, either animal or vegetable, had the remotest connection. I hold in my hand a little book, published in 1874. It gives, in brief, the substance of my earliest lectures, delivered away back in the '60's, in an effort to prove that in all geologic time the earth had an annular or ring system, such as the planet Saturn has to-day. That the progressive collapse of that system made all the "ages" and all the "deluges" the earth ever saw. Also, that the deluge of Noah, coming from that source, closed the grand drama of geologic revulsions.

With this as our basic thought, I want us to go back in imagination to the very morn of geologic time, to see how the earth got its rings, and what they were made of. We will make the attempt, and my word for it now, we will find that those rings were made out of the fiery exhalations that went up to the skies from the molten earth. We know what those exhalations were, for we know what the earth's elements are to-day, and her compounds, too. We know the oceans, vaporized, were there, and all else of the molten planet that inveterate heat could vaporize and sublime, and force into chemical activity and union. We know, too, that these fiery sublimations went to the terrestrial skies, and I want to show you that it was just as natural and inevitable for an ocean of oily hydro-carbons to go there as an ocean of water, and we all know that all the terrestrial waters were driven aloft when the earth was rocked in its cradle of flame. Let us see.

In the first place, I do not oppose the claim that organic decomposition, by the aid of heat, is competent to form oily compounds. I do not deny that fishes and crustaceans can be treated in the chemist's retort and made to form hydro-carbons, for that is nature's process when the necessary elements are at hand for the work, whether in the crucible of the chemist or in the world furnace of the molten era. I simply oppose the conclusion that geologists have drawn from this experiment. That fishes may be thus made to form oily products is no testimony to the claim that the inhabitants of the devonian and carboniferous oceans made the vast fund of petroleum, or any part of it. For it is a fact well known to the chemists that apples and pumpkins, as well as oysters, may be made to yield hydro-carbons, and are we to draw these innocent organisms from their legitimate field? It makes no difference whether the hydrogen and carbon involved in the great world process of oil-making came from animal or mineral forms.

Now, we know that there was an immeasurable amount of hydrogen and carbon in the molten earth, and we know they had no organic source. They came from the great unknown source of elements, and the Great Chemist had them in His retort,—the igneous earth,—and His fires raged about them for millions of years. In the name of law and reason, how could He fail to make them combine in that Titanic furnace, and be thus compelled in after ages to call in the puny aid of the fish? If we are to believe the prevailing geologic proposition, it was certainly a grand opportunity lost.

Let us watch the gas-maker and learn a lesson in philosophic world-making as he fills his retort and starts his fires. He may fill it with coal, or wood, paper, hair, or even with fish. With a moderate heat he drives out the watery elements, and we see a jet of steam coming out as the first product. In a short time, as the heat is increased, this jet of steam gives place to one of smoke. This is a form of carbon, and it is readily oxidized in the air and soon becomes invisible. It is burnt up; and here we learn that smoke is unburnt fuel. We also learn from analogy that if the molten earth was a smoking world, it was God's mighty fuel-former, before a bud or plant or an animal existed on the globe. And here it is impossible to avoid the conclusion that beyond a doubt a great fund of unconsumed carbon went from the molten earth to the skies, unless it was caught in an ocean of oxygen and burnt up, and this was hardly possible. But let us watch our retort. By raising the temperature further, the jet becomes a light hydro-carbon, plus carbon, an impure illuminant. A still higher temperature expels a heavier hydro-carbon, and, as the heat becomes intense, oily hydro-carbons supervene. If, now, a jet of superheated steam be forced into the retort, all these products are increased and enriched, and when the fires are allowed to cool, and the retort opened, we find a tarry form of carbon as a residuum, mingled with asphaltic or graphitic products,—all this done by the puny fires of the chemist. Compared with those of the molten earth, we place the infinitely small beside the immeasurably great, and ask what did the great world-retort expel from its boiling and surging entrails?

Now, there are some things known in the proposition we have in hand. It is known that this earth, in the dawn of geologic time, was an igneous, incandescent mass; and, whether we choose to call it the Great Chemist's crucible, a flaming sun, or scintillating star, it is all one in the grand scheme of world making. Fire held dynamic control. It is known that carbon and hydrogen were two all-abounding elements in that primitive furnace. It is known that carbon and hydrogen, thus conditioned, actively seek combination, and unless they passed through a sea of free oxygen on their way to the skies, they arose as oily products of the infant earth and filled the surrounding heavens,—light carbons, heavy carbons, asphaltic and graphitic carbons; and we know, too, that all this occurred long, long before the day of fishes. It is known that the vaporized oceans were there, a world of superheated steam, and took an active part in this plan of world evolution, ever active and eager to increase and enrich the planet's oily products. It is known that the resolution and decomposition of world matter in its primitive stage is not different from that of matter in its secondary condition, except in degree of competency; hence, if the decomposition of organic matter can make petroleum in infinitesimal quantities by bringing nascent carbon and hydrogen into contact, how much more must have been produced when all the hydrogen and carbon of the molten earth came in contact for millions of years, under conditions a thousandfold more adequate to effect rapid combination? It is, therefore, not so much a question as to the ability of the igneous earth to make oily compounds, as to how it could have failed to make them. It would be just as reasonable to deny the adequacy of the chemist's retort as that of the molten earth, since the selfsame elements are treated in the selfsame way—comparing the small with the great.

One of the great lessons we learn at the retort is, that it requires a great heat and the presence of steam to make true oily hydro-carbons, even with organic matter supplied. A molten world supplied inveterate heat and all the elements needed, and the chemist can only imitate in the most impractical way what nature is continually doing in millions of molten orbs. If the geologist denies this universal process he must also deny that hydrogen and carbon are universe elements, and so far as our world is concerned it cannot be denied, and hence he cannot for a moment logically or reasonably oppose the claim I have made that all the petroleum of the earth was found in the world furnace when it shone out as a star.

I mention these things to show you that the claims I put forth in regard to the primitive igneous origin of petroleum stand on a foundation firm and eternal. Law is its basic rock, for everywhere we see molten orbs, and as surely as carbon and hydrogen exist with them they must combine. It is hardly possible that conditions can exist on such flaming orbs to prevent their union, and we know their union means the formation of oily compounds. The only objection that can be urged against the thought is the possible surplus of oxygen in such fiery centers. But, so far as the earth is concerned, that objection is forever brushed aside by the fact that there was not enough oxygen present to saturate other elements and burn up other compounds more eagerly sought by it. It united with an immeasurable sea of hydrogen to form the waters of the earth—the vast oceans that now roll around it. The scientist actually measures the free oxygen of the molten globe by the aid of known conditions, and we cannot but conclude that the primitive hydro-carbons were not consumed. They are not consumed to-day in the hottest furnace, unless a blast of oxygen is forced through it. Millions of fires, coke-ovens and furnaces are sending up carbons and hydro-carbons all the time, and we find these fire-formed products lining every chimney and smokestack,—sooty hydro-carbons that take fire, sometimes, and burn—and we may be sure that the burning earth did the same thing, for free oxygen could not have had access everywhere into the recesses of the smoking globe.

This great problem has other known conditions. We all know that the rock formation in which oil is found to-day was formed at the bottom of the sea by sedimentation—by the depositing of matter which floated in the ancient seas. This being the case, the hydro-carbon matter that settled in the forming bed also floated in the sea at the same time. It is also well known that in many places the oil-bearing rock is utterly destitute of animal remains. Over wide areas not a fossil of a fish or crustacean has been found, and, what is most apparent, no part of those beds contain animal remains in quantities sufficient to make even a show of oil. In the same field the oil varies in specific gravity, as well as in quality. When an oil-bearing stratum is struck at a moderate depth, as a rule the oil has less specific gravity than that found in the same bed at a greater depth. The advocates of the fish cannot explain this. Did the different kinds of fish make the different kinds of oil? On the supposition that these carbons, the products of different degrees of heat, floated in the sea, it is easily explained, for the heavier carbons floated into deeper water and there settled, while the lighter floated higher and settled on higher ground. Thus the waters necessarily assorted the different grades of carbon before they settled, and thus they are separated to-day. Supposing a fish should become buried and heat and pressure should begin the work of decomposition and oil-making, the light and the heavy carbons become locked down together, on the spot, with no possible chance for them to become separated.

My friends, we have seen a world rocked in its primitive cradle of fire; we have learned that its waters were all formed in inveterate flames by the union of hydrogen with oxygen, thus disposing with so much of the latter that the vast amount of carbon escaped the devourer, and remains unconsumed and locked up in the earth's crust. As all these compounds were driven to the skies, we want to know by what process they came back to the earth. They were made away back in the igneous era, and were stored up in the silurian and later beds, showing that they remained in the telluric heavens for many millions of years after the earth cooled down. This being the case, we are forced to admit that they revolved about the earth, as an annular or ring system; for, unless they did revolve thus, they would have fallen as the earth grew cold. But we know the lapse of time between the molten era and silurian time was immeasurably vast, and it marks the long interval during which the carbons rode the skies. Now, they could not have remained on high any more than a stone, unless they revolved about the earth, and this means ring formation, for it has been practically determined that rotating vapors naturally assume the ring form. Thus, it would seem that the molten condition of a planet is the first step to ring conditions, for it is then that world vapors go to the skies, until a vast world envelope of aqueous mineral and metallic exhalations is formed. It is simply impossible that such an envelope would not form, and we know that the earth envelope contained hydro-carbons to an immeasurable amount. We know, too, that the whole fiery mass rotated. This gave great momentum to the perimeter, far in excess of that of the central mass. The perimeter of a rotating wheel moves faster than the hub. This being the case, the great rim of the primitive earth could not fall as the latter grew cold. But it would fall in after times and form stratum after stratum on the earth. For mechanical and philosophic reasons, this great earth-appendage must have fallen in grand instalments, like so many dust clouds, and was carried to the oceans, and borne by currents to different parts of the world. During all the time the carbons rode on high they were necessarily associated with mineral and watery distillations, and when they fell they were still thus associated and came down as vast deluges of muddy waters. Carried to the seas, they sank together, and together they formed oil-bearing beds. Many times I have seen oil from wells flow down and float on the surface of the Ohio River. During high water the mud particles of the stream united with the oil and fell to the bottom of the river, forming a thin stratum of oil-bearing ooze. Let us imagine a world cloud of oily particles thus carried down from on high in deluges of muddy rains. Inevitably they would go to the swamps, lakes, seas and oceans of the earth, and form oil-bearing beds. Imagine such a muddy mass of waters carried down the Mississippi and deposited in the Gulf of Mexico, mingling with the beds now forming on its bottom, and you can form some idea of how oil beds were made of fiery sublimations, first formed in the molten earth and driven to the skies, and thence in after ages returning to the surface of the planet. Imagine the Gulf Stream of the Atlantic as it issues from the Gulf of Mexico laden with this oil-steeped matter, and you can see how this matter, borne to distant parts of the ocean, would become a part of strata now forming there.

Thus, the oily carbons, formed in the world-alembic of the Great Chemist millions of years before a planet or a fish existed on the earth, arose together with other fiery distillations; together they revolved for long ages while the planet was cooling; together they fell and became a part of the world's strata, and together they lay in store for the use of man.

This being the process by which the earth became stored with the oily carbons, it is plain that the geologist, to be able to be of any practical use in locating oil fields, must familiarize himself as far as possible with the currents of the ancient seas, and forever discard the fish.

But enough of that. You want me to tell you some facts—some annular facts about the oil field of Southern California. Well, I came to Pasadena in 1887, and in the summer of 1888 I spent much of the time in search for oil indications, for I had become convinced that a vast amount of oily carbon must have been carried from the northward into the ancient Pacific Ocean, and you all know that if that carbon fell in polar lands (and it did fall there), it must have been floated into that ocean. Well, the great currents from the west and southwest, it seemed to me, must have carried that carbon right toward the coast of Southern California. But where was the coast then? The waters of the Pacific at that time dashed against the feet of the Sierras in a great semi-circular curve, extending from the Santa Barbara coast to San Bernardino, thence southward by the San Jacinto Mountains. You can see that much of the lowland, even the low hills of Ventura, Los Angeles, and nearly all of Orange County, and great part of San Diego County were at the bottom of the sea, into which the ocean currents rushed with their charges of oily carbons. I spent a great deal of thought on this feature of the case, and became convinced before I came to California that it was oil territory. I reasoned that all the geological conditions were favorable for the formation of oil beds in the waters of this great semi-circular gulf, and therefore it could hardly be possible that they were not formed, for we must remember that our theory makes oil a world-deposit, and must have been formed wherever currents favored it.

I searched in fullest confidence that I was in an oil field of vast extent. In both Santa Barbara and Ventura Counties I found oil, not only as it trickled from the rock, but far out on the surface of the ocean. This was all sufficient to vindicate the theory; for a single instance of oil thus found was competent to prove that our hypothetic currents had been at work, and if they had been thus at work on the coast of that region, it was almost certain that they had operated over the remainder of the gulf, and more effectively. From Santa Paula a search was made eastward, and a visit to the oil-steeped soil on Temple Street, Los Angeles, proved still further that an oil field of vast extent existed here. I then went to the Puente Hills, and on to the hills north of Fullerton, and found all the evidence I wanted to vindicate the claim that all the region from the mountains to the sea, and certainly extending far out into the ocean, was a great deposit of petroleum.

Having made these discoveries, I went before the Chamber of Commerce in Los Angeles and tried to move that body to institute an effort to awaken an interest in them. But the great boom was collapsing, and the calm after the tempest made my visit a failure. I told that body that petroleum was an ancient sea deposit, and the very fact that oil oozed from the hills was proof that oil-bearing rock underlaid all the region from the hills to the sea, and that the day would come when Southern California would prove to be one of the greatest oil regions on the continent. The upheaval of the hills on the south of the San Gabriel Valley has simply revealed the universal bed of hydro-carbons that the ancient currents carried to this region. That sea in which the oily carbons floated was a vast one, and hence the oily deposit is a vast one, too. We can easily bound it on the north, but on the east, south and southwest I cannot define its limits. I have repeated in a hundred lectures that the field must be very rich toward the south, but more and more barren toward the primary mountains. That wells could be sunk far out in the ocean with almost a certainty of finding oil—why not? The ocean that received the oil carbons from the north world must have been almost a boundless one. The same ocean rolled its carbon-laden waters around the San Jacinto Mountains, over the Salton basin and great Colorado Valley, and must have deposited a great fund of petroleum there. The same ocean and same current swept over the plains of Western and Southern Arizona, and must have planted some of it there. The same ocean penetrated the inland valleys of a great part of all the Pacific States, and in all these, wherever currents from the west could enter, oil certainly can be found, and even up to the Arctic Circle.

In conclusion, let me add: In prospecting for oil, keep away from the primary beds—shun the mountains. They were the ancient shore line, and petroleum could hardly settle in such shallow waters. The hills in which our oil is found are geologically modern upheavals. If you find petroleum or asphalt in one of those hills, it is almost certain that all the hills in that neighborhood are oil-bearing. In a vast ocean carrying primitive carbons, the deposit could hardly be local. A trace of asphalt or bitumen simply shows the track of old ocean bearing primitive hydro-carbons, and unless it was of local waters, it could hardly be a local oil deposit. Of course, there are many local deposits, and even in a sea laden with carbons there are necessarily barren spots, for, as I have said before, these deposits depend altogether upon the caprice of currents. A vast ocean with a vast world-current seems to have governed the formation of our petroleum beds. Fortunate California!

Can you contemplate these things and put any reliance whatever on the claim that petroleum had an animal origin? Fishes were not subject to the course of currents as primitive distillates were, and could not make a barren spot in the very midst of an oil field. In fact, the old theory of the animal origin of oil as it is found in the earth's crust has no support whatever in fact. The idea is puerile and cannot stand the light of intelligent progress.

Granting that the world was molten, these fiery distillates follow as a matter of course. Then the earth's ring system, made out of aqueous, mineral and metallic vapors, follows as a necessity. Then comes their progressive decline in the flight of ages, and their deposit here, there and everywhere that oceans rolled and currents presided. The first thing the oil prospector should do is to cast the "fish story" into the scientific dump heap and study the amazing scheme of annular world evolution.

LORD KELVIN ON FREE OXYGEN.

In a recent lecture by Lord Kelvin, that learned man in speaking of the vegetable origin of coal, said: "When the earth began to cool it was surrounded by an atmosphere of nitrogen and carbonic acid gases *without any free oxygen.*" (Italics mine.)

It is fitting that I close this volume with a few comments upon this startling announcement. If there was no free oxygen when the earth *began* to cool it was certainly very scarce for a vast length of time *before* it began to cool. This, added to the millions of years during which it was *cooling,* gives us an immeasurable period when the molten earth had no free oxygen; for, according to this learned physicist and his compeers, there is no source of free oxygen other than vegetation.

Standing upon this ground, it must be an uncertain task to determine in what part of archæan time there was enough free oxygen present to burn up the carbon in the molten earth. It could not have been burnt up during the cooling period, nor in the vast eon preceding it; for it takes free oxygen to burn carbon, and hence it must have been devoured in the pre-molten era. But how and where did that era get its free oxygen? Will Lord Kelvin supply it from pre-archæan vegetation?

Seriously now, we are compelled to admit that a vast amount of the earth's primitive carbon was not consumed at all; any more than if it had been placed in a sealed retort and subjected to inveterate heat. Even if we have to concede that some time during the earth's greatest igneous activity free oxygen was present we cannot for a moment admit that it was so plentiful and so vigilant that no carbon exhalations rising from the seething planet could escape unconsumed into the skies. We cannot accomplish this even by driving a current of pure free oxygen through our blast furnaces; how much less effectively would free oxygen act in the molten era, when trammeled by the presence of an atmosphere of fiery elements having a more eager appetite for oxygen, than carbon has?

Granting that the atmosphere of the molten earth was rich in carbonic acid (free oxygen united with carbon), where was the vegetation that supplied the oxygen, if that be its only source? The simple fact is that it makes no difference what the ancient or modern source of oxygen was, or is; it could not, as law is supreme, consume the carbon of the molten earth. We see great beds of nearly pure iron and other metals left unconsumed, and these we know rapidly deflagrate when heated in free oxygen. If these metals escaped, carbon could not fail to be gathered from the earth's inmost bosom and be borne as an unconsumed fuel to the loftiest heights of the primitive atmosphere.

Especially must this have been the case if "when the earth began to cool it was surrounded by an atmosphere of nitrogen and carbonic acid gases, without any free oxygen." It is amazing to what unnatural conclusions the vegetation theory leads the scientist. The day is coming when some stronger head than this will meet these ironclad exponents of a false philosophy, and show to a laughing world what an illegitimate birth modern geology is. God speed the day!

LIST OF THE AUTHOR'S PUBLICATIONS.

1. "The Earth's Annular System," 4th edition; 407 pages, illustrated, cloth. Price, $2.50.
2. "The Misread Record," (The Deluge) 130 pages, cloth. Price, $1.00. Paper, 65 cents.
3. "Alaska, Land of Nuggets, Why?" 68 pages, cloth. Price $1.00, paper, 50 cents.

PAMPHLETS.

4. "Eden's Flaming Sword. What Was It?" 48 pages. Price, 25 cents.
5. "The Coal Problem." (Origin of Coal), 48 pages. Price, 25 cents.
6. "The Lost Lake" (Geological Problem), 40 pages. Price, 25 cents.
7. "Ophir's Golden Wedge," (A Polar Puzzle). 40 pages. Price, 25 cents.
8. "Waters Above the Firmament." (Souvenir reprint of the author's first lecture on the Annular Theory and first published in 1874, mailed free to all Annular-Canopy Students).

"The Great Red Dragon," 2d edition. Manuscript, about 150 pages; to be published when there is sufficient demand. The Dragon of all mythologies is here shown to have been the earth-canopy personified.

These works can be had by addressing the

ANNULAR WORLD COMPANY,
411 Kensington Place,
Pasadena, Cal.

Spotted something that isn't quite right?

Let us know and we'll fix it for future reprints. And as a thank you for your diligence, we'll send you a free Sothis book of your choice.

publisher@sothis.press

This page intentionally left blank

Made in the USA
Las Vegas, NV
17 April 2026

45524479R10252